VENKATESHA PERUMAL

INDUSTRIAL BOILERS

Principles, Design, and Operation A Comprehensive Guide for Power & Process Professionals

First edition

Contents

Preface

Walk into any industrial plant in India—a sugar mill in Maharashtra, a chemical plant in Gujarat, a power station in Tamil Nadu, or a textile factory in Coimbatore—and you'll find one constant: the boiler room.

It's not just another equipment room. It's the beating heart of the operation. Stop the boiler, and everything stops.

I've spent over 10 years commissioning power plants and troubleshooting boiler systems across India. I've been the engineer called at 2 AM when something goes wrong. I've seen boilers run beautifully for decades, and I've investigated failures that cost companies crores and, tragically, sometimes lives.

Every lesson in this book comes from real experience—the kind you don't find in textbooks alone.

Who This Book Is For

If you're a STUDENT preparing for mechanical or chemical engineering exams, you'll find clear explanations connecting thermodynamic theory to actual boiler operation. No more memorizing formulas without understanding why they matter or where they're used.

If you're an OPERATOR preparing for your Boiler Attendant Certificate (First Class or Second Class), you'll get practical knowledge for both passing the exam and safely running a real boiler. Chapter 9 contains exam-style questions based on actual certification tests.

If you're an EXPERIENCED ENGINEER managing boiler operations, you'll find detailed technical content on water chemistry programs, efficiency optimization, failure analysis, and regulatory compliance—with thorough coverage of both ASME standards and Indian Boiler Regulations (IBR).

A Critical Word on Safety

Throughout this book, you'll notice I emphasize safety repeatedly. That's not by accident.

I've investigated boiler failures. I've seen the aftermath. Every safety warning in these pages represents a real incident—often with devastating consequences.

A boiler is a pressure vessel containing thousands of kilograms of water superheated beyond its normal boiling point, held under enormous pressure in a steel container. The stored energy in a typical 10 TPH boiler at 10 bar pressure is equivalent to several hundred kilograms of TNT.

Respect it. The ASME code and IBR regulations weren't written arbitrarily—they exist because people died learning these lessons. Every safety device, every inspection requirement, every operational procedure exists for a reason.

Never bypass a safety device. Never defer maintenance. Never take shortcuts. Your life and the lives of your coworkers depend on it.

Acknowledgments

This book wouldn't exist without the hundreds of operators, maintenance engineers, and plant managers I've worked with over the years. Their questions, their challenges, their hard-won insights have shaped every chapter.

Special thanks to the operators who've patiently explained 'how things really work' versus 'how the manual says they work.' That gap between theory and practice is what this book tries to bridge.

One Final Note

Technical knowledge alone doesn't make a good boiler operator. You also need judgment, patience, attention to detail, and respect for the equipment.

You need to develop the instinct that tells you when something 'doesn't sound right' or 'doesn't look normal'—even before your gauges show a problem. That instinct comes from experience, but this book will help you build the foundation faster and more safely.

Stay safe. Operate smart. Learn continuously.

Let's begin.

Venkatesha Perumal Ramasamy

Prologue

THE MIDNIGHT CALL FROM CHITTAGONG

The phone rang at 11:47 PM on a Tuesday night. I was in my apartment in Chittagong, reviewing water analysis reports from the day shift.

It was Shahid, my plant manager, calling from the chemical plant. His voice was tense but controlled—the kind of calm that comes from years of handling emergencies.

"We have a situation with Boiler #2. I'm at the plant now. How fast can you get here?"

I was the Water Treatment Plant Incharge at the time, responsible for ensuring the boiler feedwater met specifications. "What kind of situation?"

"The operators noticed something wrong about an hour ago. Steam quality dropping. Pressure fluctuating. Foaming in the sight glass." He paused. "I've ordered a controlled shutdown. We need you here to figure out what happened before we drain and inspect."

"I'll be there in 20 minutes."

I grabbed my laptop with the water treatment logs, my testing kit, and headed to the plant. During the drive through Chittagong's quiet streets, I ran through possibilities in my mind.

Foaming. Pressure fluctuations. Dropping steam quality.

Those symptoms pointed to water chemistry problems. But we'd been running stable for weeks. What changed?

* * *

When I arrived, Shahid was in the control room with the shift supervisor and chief operator. The boiler was cooling down, pressure dropping gradually from 10 bar to 5 bar. They'd done it by the book—no thermal shock, no cold water injection.

"Show me the logs," I said.

The operator handed me the shift logbook. Water level readings normal. Feedwater temperature normal. Chemical dosing pumps...

Wait.

"When did you last check the chemical dosing tank?" I asked.

The operator looked uncomfortable. "This morning, sir. Why?"

I walked to the water treatment room with Shahid following. The chemical dosing system had three tanks: one for phosphate (prevents scale), one for sulfite (removes oxygen), and one for caustic (controls pH).

I checked the phosphate tank. Nearly empty—that was normal, we were due for a refill tomorrow.

I checked the sulfite tank. Full. Strange—it should have been half-empty by now.

Then I checked the caustic tank. Also nearly empty.

I looked at the piping. My heart sank.

Someone had connected the dosing pump to the wrong tank. Instead of dosing 30-40 ppm of phosphate into the boiler, we'd been pumping 500+ ppm of caustic soda for the past 8 hours.

Extremely high caustic in boiler water causes caustic

carryover—exactly what creates foaming, unstable water levels, and poor steam quality.

"We need to drain the boiler completely," I told Shahid. "We can't restart until we've flushed this out."

* * *

At 2 AM, with the boiler cooled to 80°C and depressurized, we began the drain. The water coming out was cloudy—bad sign. Normal boiler water should be clear.

I collected samples and ran quick tests:

- pH: 13.2 (should be 10.5-11.5)
- Caustic: 1,850 ppm (should be <500 ppm)
- Conductivity: 8,500 μS/cm (should be <3,000 μS/cm)

The numbers confirmed it. Massive chemical overdose.

"How did this happen?" Shahid asked.

We found out the next morning. A new maintenance worker had disconnected the chemical dosing lines during a routine pump maintenance. When reconnecting them, he'd swapped the phosphate and caustic lines—they were the same diameter pipe, and he didn't check the labels carefully.

The day shift operator had noticed the sulfite tank wasn't going down, but assumed the pump had air in it and would catch up. He didn't think to check if chemicals were actually entering the boiler—or from which tank.

* * *

We spent the entire night draining, flushing, and refilling the boiler with fresh treated water. Internal inspection the next day showed no permanent damage, fortunately. The tubes were clean, no caustic corrosion yet—we'd caught it in time.

The plant lost 36 hours of production. The cost in lost output, emergency labor, and chemical waste exceeded ₹8 lakhs (about 9 lakh Bangladeshi Taka at the time).

But it could have been so much worse. If we'd continued running, the high caustic could have caused:

- Caustic stress corrosion cracking (tube failures)
- Carryover contaminating the product (pharmaceutical plant—entire batches ruined)
- Turbine blade deposits (if we'd had turbines)

Shahid handled the emergency perfectly—recognized the symptoms, shut down safely, called the right people. His experience prevented a minor problem from becoming a catastrophe.

* * *

That night taught me several lessons that shaped my entire approach to boiler water treatment:

First: Chemistry problems show symptoms before they cause failures. Foaming, unstable levels, poor steam quality—these are warnings. Listen to them.

Second: Even experienced operators need to understand water chemistry, not just follow procedures. If the day shift had understood why sulfite consumption matters, they would have investigated immediately.

Third: Simple mistakes (wrong pipe connection) can have major consequences. Double-check everything. Label everything clearly. Train everyone thoroughly.

Fourth: You need people who know when to shut down. Production pressure is real, but Shahid didn't hesitate when he saw abnormal conditions. That decision saved the plant.

* * *

I wrote this book because I don't want others to make the same mistakes we made in Chittagong that night.

I wrote it for three people:

For the student who's learning about boilers in a classroom but needs to understand what these concepts mean in a real plant at midnight when something goes wrong.

For the operator who runs the boiler every day and needs to recognize when something isn't right—even if all the instruments say everything is fine.

For the engineer managing water treatment, operations, or maintenance, who needs to understand the chemistry, the equipment, the procedures, and how they all connect.

Every chapter in this book addresses real problems I've encountered in over 10 years working with boilers across India and Bangladesh. Every warning represents something that actually went wrong somewhere. Every procedure exists because we learned—sometimes painfully—why it matters.

This book covers both ASME international standards and Indian Boiler Regulations (IBR), uses costs in rupees, references Indian and regional manufacturers, and addresses challenges specific to our industries—from monsoon water quality changes to the reality of working with limited resources.

If you work with boilers, or plan to, this book might prevent your own midnight emergency call.

Pay attention to the water chemistry chapters—they're not just theory. Understand the safety procedures—they exist for reasons. Learn the troubleshooting methods—you'll need them.

And if you ever notice something that doesn't seem right, even if you can't explain exactly what's wrong—speak up. Trust that instinct. It's better to investigate and find nothing than to ignore a warning sign.

Water treatment is the foundation of good boiler operation. Chemical dosing seems simple—pump A into tank B—but get it wrong and you can damage a multi-crore asset in hours.

A boiler is a partnership between mechanical systems and chemical processes. Both matter equally.

Let's make sure you understand both.

Introduction

What Makes This Book Different

Indian Industrial Context: All examples use Indian costs (in rupees, not dollars). We reference Indian manufacturers (Thermax, Forbes Marshall, Triveni). We address challenges specific to Indian industry: monsoon water quality variations, regional fuel availability (coal, bagasse, rice husk), and IBR compliance requirements.

Dual Regulatory Coverage: Most international books cover only ASME. Most Indian books cover only IBR. This book thoroughly covers BOTH, because if you operate a boiler in India, you need to comply with international design standards AND Indian statutory regulations.

Practical, Not Just Academic: I've written this the way I explain things to junior engineers on-site—clearly, with real examples, respecting your intelligence. No unnecessary jargon. No complex language just to sound impressive.

Complete Technical Reference: You'll find actual steam tables, water chemistry targets by pressure range, and conversion factors you can use on the job—not just references to 'consult other books.'

How to Use This Book

- **If you're new to boilers:** Start with Chapter 1 and work through sequentially. Each chapter builds on previous ones.
- **If you're studying for an exam:** Focus on Chapters 4-7 (water treatment, chemistry, operation, troubleshooting) and work through Chapter 9's practice questions.
- **If you're improving plant efficiency:** Jump to Chapter 8, but understand you'll need good water chemistry (Chapters 4-5) to achieve and maintain good efficiency.
- **If you're investigating a problem:** Chapter 7 (Troubleshooting) is organized by symptom and failure type for quick reference.

I

Part One

1. Introduction - The Heart of the Industrial Plant

1.1 Why Steam Still Powers Modern Industry

Let me start with a simple question: In 2026, with all our advanced technology—solar panels, batteries, computer control systems—why do we still boil water to run factories?

The answer isn't nostalgia. It's physics and economics.

Steam remains the most efficient way to move large amounts of thermal energy from point A to point B in an industrial setting. Nothing else comes close for the cost and scale required.

Think about a typical chemical plant or power station. You need to:

- Heat chemical reactors to precise temperatures
- Run turbines to generate electricity
- Sterilize pharmaceutical equipment
- Heat buildings in winter
- Power dozens of different processes simultaneously

You can't run electric heaters everywhere (the power bill would bankrupt you). You can't burn fuel at each location (too dangerous and inefficient). But you CAN generate steam in one central boiler house and pipe it anywhere you need it at whatever pressure and temperature the job requires.

The Magic of Water

Water is special. Really special. Here's why engineers love it:

- **High Heat Capacity:** Water can absorb enormous amounts of heat before its temperature rises much. Technically, it takes 4.18 kJ to raise 1 kg of water by just 1°C. That's way higher than most liquids. This means water is great at storing and transporting heat energy.
- **Massive Latent Heat:** But here's where it gets interesting. When water changes from liquid to steam, it absorbs 2,257 kJ/kg WITHOUT getting any hotter. Read that again—over 2,000 kilojoules absorbed with zero temperature change. This hidden energy is called the "latent heat of evaporation."

What does this mean practically? You can pack a huge amount of energy into steam, pipe it across your entire facility, and when it condenses back to water, it releases all that energy instantly and precisely where you need it.

I remember commissioning a textile dyeing plant where they needed to maintain 95°C ±2°C in huge dye vats. Electric heating couldn't give them that precision—temperature would swing wildly. Steam at 1.5 bar (109°C saturation temperature) gave them perfect control. When the vat needed heat, steam condensed. When it was hot enough, condensation stopped automatically. No complex controls required—just physics.

The Pressure-Temperature Relationship

Here's something that confuses newcomers: Why do we run boilers at high pressure?

At normal atmospheric pressure (1 bar), water boils at 100°C. If you keep adding heat, the water doesn't get hotter—it just boils faster, turning to steam.

But—and this is critical—if you INCREASE the pressure, the boiling point rises:

- At 5 bar: water boils at 152°C
- At 10 bar: water boils at 180°C
- At 40 bar: water boils at 250°C
- At 100 bar: water boils at 311°C

This means higher pressure = higher temperature = more energy per kilogram of steam.

That's why power plants run at incredibly high pressures (160+ bar). More pressure means more energy means more electricity from the same amount of fuel.

Think of your kitchen pressure cooker. Food cooks faster not just because of pressure, but because the water inside is HOTTER than 100°C (maybe 120°C at 2 bar). Same principle, just scaled up a million times.

Real-World Steam Economics

Let me give you actual numbers from a mid-sized chemical plant I worked with:

Option 1: Individual electric heaters at each process

- Total heating load: 5,000 kW
- Electricity cost: ₹7/kWh (industrial rate)
- Monthly cost: 5,000 kW × 720 hours × ₹7 = ₹25.2 lakhs

Option 2: Central boiler generating steam

- 10 TPH boiler running on coal
- Coal consumption: 1.3 kg/kg steam × 10,000 kg/h = 13,000 kg/h
- Coal cost: ₹6/kg
- Monthly cost: 13,000 kg/h × 720 hours × ₹6/kg = ₹5.6 lakhs

Savings: ₹19.6 lakhs per month. That's why steam still rules.

1.2 Classifying Boilers: Legal and Technical Categories

In the professional world, we don't just call them "big boilers" or "small boilers." We classify them by legal requirements and design architecture, because the rules change dramatically based on what you're operating.

Legal Classifications (ASME Code)

The American Society of Mechanical Engineers (ASME) created the Boiler and Pressure Vessel Code after thousands of people died in the 1800s and early 1900s from boiler explosions. This code is now the global standard, including in India.

Power Boilers (ASME Section I):

- Generate steam for power generation or process use
- Operating pressure typically >15 psi (1 bar)
- Strict material, design, welding, and inspection requirements
- Must have the "S" stamp on the nameplate
- Requires qualified operators (in India: Boiler Attendant Certificate)

Example: A 15 TPH boiler at 20 bar (290 psi) generating

steam for a chemical plant process. This is Section I.

Heating Boilers (ASME Section IV):

- For space heating and hot water only
- Low pressure: <160 psi (11 bar)
- Low temperature: <250°F (121°C)
- Less strict requirements because stored energy is much lower

Example: A building heating boiler for a shopping mall. This is Section IV.

Why does this matter? Insurance, permits, inspection frequency, and operator requirements are completely different. A Section I boiler might need weekly operator checks and annual inspections by government inspectors. A Section IV boiler might only need annual service.

Table 1.1: American Society of Mechanical Engineers (ASME)

Classification	Definition (ASME BPVC)	Typical Applications
Low-Pressure (Heating)	Steam < 15 psig. Hot water < 250°F and < 160 psig.	Schools, hospitals, office space heating, domestic hot water.
High-Pressure (Power)	Steam > 15 psig. Hot water > 250°F or > 160 psig.	Electrical utilities, refineries, chemical processing, paper mills.

Design Classifications

Fire-Tube Boilers:

Hot combustion gases flow INSIDE tubes, water surrounds the OUTSIDE of tubes.Think of it like this: imagine a cylindrical water tank with dozens of metal pipes running through it. Fire and hot gases flow through these pipes, heating the water surrounding them.

Characteristics:

- Pressure limit: typically <25 bar (363 psi)
- Capacity limit: typically <25 TPH
- Compact design
- Easy to maintain
- Lower capital cost
- Startup time: 2-4 hours

Common uses: Small to medium industrial applications, hotels, laundries, food processingI've worked with dozens of fire-tube boilers. They're workhorses—forgiving, reliable, and hard to damage if you follow basic operating rules. Perfect for applications that don't need extremely high pressure.

Water-Tube Boilers:

Water flows INSIDE tubes, hot gases surround the OUTSIDE of tubes.

Imagine hundreds of vertical tubes filled with water, with fire and hot gases flowing around them on the outside.

Characteristics:

- Pressure: can exceed 300 bar (4,350 psi)
- Capacity: no practical limit (can be >1,000 TPH)
- Large footprint
- More complex maintenance

- Higher capital cost
- Startup time: 4-12 hours
- Higher efficiency at high pressures

Common uses: Power generation, large industrial processes, any application requiring high pressure or large capacity
Water-tube boilers are like Formula 1 race cars—higher performance, but demand more attention and expertise.

Table 1.2: Fire-Tube vs Water-Tube Detailed Comparison

Feature	Fire-Tube Boiler	Water-Tube Boiler
Max Pressure	25 bar (363 psi)	300+ bar (4,350+ psi)
Max Capacity	25 TPH typical	Unlimited (1,000+ TPH possible)
Physical Size	Compact, horizontal	Large, vertical or horizontal
Water Volume	Large (slow response)	Small (fast response)
Startup Time	2-4 hours	4-12 hours
Efficiency	75-82%	85-92%
Maintenance Access	Easy	Complex
Capital Cost (₹/TPH)	₹30-50 lakhs	₹50-100 lakhs
Operating Cost	Moderate	Lower (better efficiency)
Footprint	Small	Large
Water Quality Needed	Moderate	Strict
Common Applications	Hotels, laundries, small industry	Power plants, large industry
Tube Replacement	Difficult	Easier
Safety	Good if maintained	Excellent

1.3 How a Boiler Actually Works: The Pressure Cooker Analogy

If you truly understand how your kitchen pressure cooker works, you understand a boiler. Let me break this down.

Scenario 1: Open Pot (Normal Cooking)

- Water boils at 100°C (at sea level, 1 bar pressure)
- No matter how much MORE heat you add, water stays at 100°C
- Extra heat just makes it boil faster and turn to steam faster
- Steam escapes into the air (wasted energy)
- Food cooks slowly because maximum temperature is 100°C

Scenario 2: Pressure Cooker (Your Kitchen)

- You seal the pot—steam can't escape
- Pressure builds as water boils and creates steam
- Boiling point rises (maybe to 120°C at 2 bar)
- Food cooks faster because it's HOTTER
- Safety valve prevents over-pressure

Scenario 3: Industrial Boiler (Scaled Up)

- Sealed steel vessel (can handle 10, 20, 100+ bar pressure)
- Water heated by fire/oil/gas burners
- Boiling point can reach 180°C, 250°C, even 350°C depending on pressure
- Steam carries massive energy to wherever it's needed
- Multiple safety systems (pressure relief valves, low-water cutoffs, etc.)

Three Types of Heat in a Boiler

1. Sensible Heat (The Temperature Rise)

This is energy needed to heat water from room temperature up to boiling point. You can "sense" it with a thermometer—the temperature goes up.

Example: Heating 1,000 kg of water from 25°C to 100°C:

Heat needed = mass × specific heat × temperature change

= 1,000 kg × 4.18 kJ/kg°C × 75°C = 313,500 kJ = 313.5 MJ

That's about 90 kWh of energy (roughly ₹630 worth of electricity at industrial rates).

2. Latent Heat of Evaporation (The Magic Heat)

Once water reaches boiling point, it stops getting hotter. ALL additional heat goes into breaking the molecular bonds between water molecules, turning liquid into steam.

Continuing our example: Converting that 1,000 kg of water at 100°C into steam at 100°C:

Heat needed = mass × latent heat

= 1,000 kg × 2,257 kJ/kg = 2,257,000 kJ = 2,257 MJ

That's 627 kWh (roughly ₹4,389 worth of electricity).

Notice: Latent heat is 7 TIMES MORE than sensible heat. This is why steam is so powerful—it stores huge amounts of hidden energy.

3. Superheat (Extra Temperature)

Some boilers (especially for power generation) take it further. After making steam, they heat it even MORE in a separate section called a superheater.

This "superheated steam" might be 400°C, 450°C, even 540°C at the same pressure. This additional heat energy improves efficiency in turbines.

Practical Example: Heating a Chemical Reactor

Let's say you need to heat a chemical reactor to 180°C for a polymerization process.

Option A: Direct fuel burner

- Dangerous (open flame near chemicals)
- Uneven heating (hot spots can damage product)
- Hard to control precisely
- Risk of fire/explosion

Option B: Electric heating jacket

- Expensive (high power consumption)
- Slow response time
- Still somewhat uneven heating

Option C: Steam jacket at 12 bar

- Steam at 12 bar saturates at 188°C
- Provides gentle, even heating
- Automatic temperature control (when reactor reaches 180°C, steam stops condensing)
- Safe (no fire risk)
- Precise and reliable

This is why chemical plants use steam. It's not just about moving heat—it's about control, safety, and precision.

1.4 Always Safety First:

We often take the safety of our plants for granted, but the history of steam is a violent one. In the 19th century, boiler explosions were so common they were considered "acts of God." In reality, they were acts of poor engineering and lack of water treatment.

The Sultana Disaster (1865):

The deadliest maritime disaster in U.S. history was not the Titanic; it was the explosion of the steamship Sultana. Carrying over 2,000 passengers (mostly Union soldiers returning from the Civil War), the ship's boilers exploded on the Mississippi River. Over 1,800 people perished. The cause? Leaking boilers, poor repairs, and "dirty" river water causing scale and overheating.

The Birth of the ASME Code:

The public outcry from thousands of such deaths led to the 1914 issuance of the ASME Boiler and Pressure Vessel Code (BPVC). This document changed the world. It standardized:

- Materials: What kind of steel can be used.
- Design: How thick a drum must be to hold a specific pressure.
- Inspection: The "Third Party" check to ensure the manufacturer didn't cut corners.

Today, if you see the "S" or "H" stamp on a boiler's nameplate, you are looking at a century of safety evolution. Operating a boiler without respecting this code is not just dangerous—it is professional negligence.

1.5 The Modern Boiler: A Chemical Reactor

As your resident chemical expert, I must add a crucial layer to this introduction: A boiler is a high-temperature chemical reactor. When you take "normal" water and put it into a boiler, you are subjecting it to temperatures that make it highly aggressive. Minerals that were invisible at room temperature (like Calcium and Silica) suddenly become hard "stone" (scale) on the tube surfaces. Oxygen, which we breathe to live, becomes a "acid-like" pit-seeker that can eat through a steel tube in weeks.

Therefore, the modern boiler operator must be part-mechanic and part-chemist. If the "Heart" of the plant is to keep beating, we must protect its "blood"—the water.

This is a significant expansion that deepens the introductory chapter, transforming it from a general overview into a professional-grade technical introduction, consistent with the tone of a comprehensive industrial handbook.

I have broken down the new information into three major, detailed sections, following the established numbering convention.

1.6 Global Governance: The ASME BPVC and the Indian Boiler Regulation (IBR)

The high-pressure world of steam cannot operate on trust; it requires mandatory, internationally recognized standards. The ASME Boiler and Pressure Vessel Code (BPVC), birthed from tragedy, serves as the most widely adopted global benchmark for mechanical integrity. However, in many key industrial markets, local statutory regulations take precedence. In India, the safety and integrity of steam generators are governed by the Indian Boiler Regulation (IBR).

1.6.1 The ASME Boiler and Pressure Vessel Code (BPVC) Sections

The BPVC is a massive compilation of over 30 volumes, representing consensus standards for the design, construction, inspection, and testing of boilers, pressure vessels, and nuclear components. For the boiler operator and engineer, a few sections are critical to daily professional life.

The operator must recognize that the "S" stamp on the boiler nameplate is a guarantee that the unit was built to these mandatory specifications. Any repair, modification, or re-rating requires strict adherence to the appropriate sections, particularly Section I and Section IX.

Table 1.3: The ASME Boiler and Pressure Vessel Code (BPVC) Sections

ASME Section	Title	Relevance to Boiler Operation
Section I	Rules for Construction of Power Boilers	This is the foundation. It dictates the minimum requirements for design, materials, and fabrication that qualify a boiler to carry the "S" stamp (Power Boiler) or the "A" stamp (Assembly). If your boiler operates above 15 psig, it must comply with this code.
Section II	Materials	This section provides the exhaustive list of approved materials (ferrous and non-ferrous) and their allowable stresses, dictating what metals can be safely used at high temperatures and pressures.
Section IV	Rules for Construction of Heating Boilers	This section covers low-pressure boilers (steam at or below 15 psig) and hot water heating boilers. It applies less stringent requirements than Section I due to the lower stored energy risk.
Section V	Nondestructive Examination (NDE)	Specifies the procedures for critical inspections, such as radiographic testing (RT), ultrasonic testing (UT), and magnetic particle inspection (MPI) used to ensure weld integrity before and during a boiler's service life.
Section IX	Welding, Brazing, and Fusing Qualifications	The definitive standard for qualifying welders and welding procedures. Boiler tubes and drums are often welded under extreme pressure conditions, meaning every joint must be provably sound.
Section X	Fiber-Reinforced Plastic Pressure Vessels	Relevant for specialized processes, though less common for high-temperature steam boilers.

1.6.2 The Indian Boiler Regulation (IBR)

The IBR is the statutory authority in India, governing the registration, inspection, and operation of all boilers and boiler components throughout the country. While sharing the core safety philosophy of ASME, the IBR has key operational differences focused on rigorous state-mandated oversight:

- **Mandatory Registration:** Every boiler and major steam pipe in India must be registered under IBR. This process is complex and tracks the life of the asset.

- **IBR-Certified Components:** All materials, valves, fittings, and pipes used in IBR-compliant systems must be certified and stamped as IBR-approved. This is a critical procurement requirement.
- **Compulsory Inspections:** IBR mandates periodic inspections by certified Boiler Inspectors (appointed by the Chief Inspectorate of Boilers in each state). No boiler can operate without a valid Certificate of Fitness to Operate issued by these authorities after a thorough hydrostatic test and internal examination.
- **Operator Competency:** IBR specifies the necessary qualifications and certifications for boiler operators and engineers, often requiring specific examinations to hold a license to operate a plant of a certain capacity.

In practical terms, the IBR is less a design code (relying on international standards for metallurgy) and more a Regulatory and Operational Code. A boiler built to ASME Section I must still comply with IBR's operational and inspection mandates when installed and commissioned in India. This dual layer of regulation ensures one of the highest safety standards globally.

1.7 The Engine of Power: The Rankine Cycle

The ultimate expression of steam's potential is the generation of electricity. The process that links the heat from the boiler to the turning of a generator is described by the Rankine Cycle—the foundational thermodynamic model of most modern thermal power plants. Understanding this cycle is understanding the purpose of high-pressure, high-temperature steam.

1.7.1 Historical Context: The Steam Revolution

The journey to the Rankine Cycle is the story of the Industrial Revolution.

- **Early Engines (1700s):** Early engines by Thomas Newcomen used steam but were incredibly wasteful. They relied on condensing steam inside the cylinder to create a vacuum, consuming massive amounts of coal.
- **James Watt's Breakthrough (1765):** Watt introduced the separate condenser. By moving the condensation step out of the main cylinder, he kept the cylinder hot, dramatically improving efficiency and paving the way for the true steam engine. This concept of continuous, separate phase change is the philosophical precursor to the Rankine Cycle.
- **William Rankine (1859):** Scottish engineer and physicist William John Macquorn Rankine formally modeled the ideal heat engine process for steam, giving the world the four-step cycle that underpins our modern power grid.

1.7.2 The Four Pillars of the Rankine Cycle

The Rankine Cycle converts heat energy into mechanical work through four idealized, continuous processes:

Isentropic Compression (The Pump):

- **Process:** Liquid water (condensate) is pumped from the low pressure of the condenser back up to the high pressure of the boiler drum.
- **Why it Matters:** This is the most energy-intensive process

in the cycle, but it is necessary because pumping a liquid requires far less energy than compressing steam. This is the first energy input.

Isobaric Heat Addition (The Boiler):

- **Process:** Water enters the boiler tubes (economizer), receives sensible heat, is boiled into saturated steam, and then often receives more heat in the superheater to become superheated steam.
- **Why it Matters:** The energy of the fuel is transferred to the working fluid (water). The goal is to maximize the heat input without damaging the metallurgy. Superheating is critical, as it raises the enthalpy and ensures the steam is dry before it enters the turbine blades.

Isentropic Expansion (The Turbine):

- **Process:** High-pressure, high-temperature steam expands across the turbine blades. This expansion forces the blades to rotate, converting the thermal and pressure energy directly into mechanical shaft work.
- **Why it Matters:** This is where the useful work (power generation) is extracted. The steam pressure drops dramatically as the energy is consumed.

Isobaric Heat Rejection (The Condenser):

- **Process:** The low-pressure steam exiting the turbine is cooled (typically by cooling water or air) and reverts entirely back into a liquid (condensate).

- **Why it Matters:** Condensing the steam creates a vacuum, maximizing the pressure differential across the turbine (step 3), which maximizes the work extracted. It also returns the working fluid to a liquid state for easy pumping (step 1).

1.7.3 Real-World Power Plant Examples

- **Subcritical Plants:** These operate below the critical point of water (3,204 psig or 22.1 MPa). They are the most common industrial and utility boilers, using clear drums and natural or assisted circulation.
- **Supercritical Plants:** To increase efficiency (the percentage of heat converted to work), modern power plants push pressure above the critical point. At this point, water transforms directly into steam without a distinct boiling phase, eliminating the heavy steam drum and improving thermal efficiency by several percentage points.
- **Combined Cycle Plants:** While not purely Rankine, these use a gas turbine (Brayton Cycle) whose exhaust heat then fuels a Heat Recovery Steam Generator (HRSG) to power a steam turbine (Rankine Cycle). This dual-cycle approach achieves the highest efficiencies in the industry.

1.8 Case Study: Low-Water Fuel Cutoff (LWFC) Failure – The Inevitable Explosion

While corrosion and scaling cause slow, insidious damage, the most immediate and catastrophic failure mode is a Low-Water Condition. This case study illustrates a common path to disaster when operational discipline fails.

Incident Profile:

- **Unit Type:** Packaged Firetube Boiler, 500 hp (Low-Pressure operation).
- **Failure Mechanism:** Acute overheating due to loss of water level.
- **Direct Cause:** Failure of the primary and secondary Low-Water Fuel Cutoff (LWFC) systems.

The Chain of Events:

- **Initial Condition:** The boiler is operating at 10 psig. A small leak develops in the feedwater line or a trap, causing a net loss of water from the boiler.
- **Control Failure:** The primary boiler level controller fails to sense the dropping level or fails to trigger the feedwater pump. The water level continues to drop past the visible glass.
- **The First Safety Barrier Fails (LWFC-Primary):** The first low-water fuel cutoff (typically a float switch or probe) is reached. This device is designed to interrupt the burner circuit immediately. In this scenario, the probe was fouled with mud/sludge due to lack of blowdown, preventing the

circuit from opening.

- **The Second Safety Barrier Fails (LWFC-Secondary):** The level continues to drop to the most critical point. The secondary LWFC—a non-resetting device—is reached. In this scenario, the operator, knowing the primary LWFC was faulty, bypassed the secondary device with a jumper wire to keep the plant running, violating safety protocol.
- **Critical Overheating:** With the water level below the highest point of the firetube sheet, the tubes are now exposed to the direct flame heat (2,000°F+) without the cooling effect of water. The temperature of the steel rises rapidly from ~375°F to well over 1,000°F, softening the metal.
- **Catastrophic Failure:** The operator notices the steam pressure dropping rapidly (due to the fire being shut down by a separate pressure switch) and rushes to add cold feedwater. As the cold water hits the superheated, softened steel of the tubes, it instantly flashes into a massive volume of steam far exceeding the boiler's design pressure. The weakened steel in the crown sheet ruptures, releasing the entire volume of stored hot water and steam.

The Lesson:

This explosion was not a metal fatigue issue; it was a procedural and chemical failure. The root cause was the fouling of the safety devices (a chemical problem) and the intentional bypassing of a safety interlock (an operational problem). Every safety device—especially the LWFC—must be tested daily and kept chemically clean. The moment a safety device is bypassed, the boiler is no longer a machine; it is a bomb.

1.9 Chapter Summary and Key Takeaways

- **ASME Code Structure:** The ASME BPVC (especially Section I for Power Boilers) provides the global mechanical standards for design, materials (Section II), and welding (Section IX).
- **IBR's Operational Focus:** The Indian Boiler Regulation (IBR) governs the mandatory registration, inspection, and certification of all boilers, adding a critical layer of statutory oversight and operator qualification that supersedes international operational standards in India.
- **Rankine Cycle:** All modern power generation relies on the four-step Rankine Cycle (Pump → Boiler → Turbine → Condenser) to efficiently convert heat into mechanical work via steam's latent heat.
- **Efficiency Drives Pressure:** Modern power generation moves toward Supercritical pressures to maximize thermodynamic efficiency by eliminating the traditional boiling phase.
- **Low-Water = Catastrophe:** The most critical operational failure is a low-water condition. The failure of primary and secondary Low-Water Fuel Cutoff (LWFC) systems, often due to chemical fouling or operator bypass, is the direct precursor to a boiler explosion. Safety systems must never be defeated.
- **Firetube vs. Watertube:** Choose Firetube for low-pressure, steady loads (The Tank). Choose Watertube for high-pressure, high-capacity, and fast response (The Spiderweb).

2. FOLLOWING THE WATER'S JOURNEY

2.1 The Steam Loop: Understanding the Cycle (Feedwater → Steam → Condensate)

The steam system cycle is fundamentally a continuous, closed-loop process designed to harvest the latent heat energy of water and return the pure water for reuse. A boiler's primary function is to serve as the phase-change device, converting liquid Feedwater (FW) into gaseous Steam.

The operational cycle is composed of three interconnected stages:

The Feedwater/Boiler Water Section (The Preparation Phase):

- **Makeup Water (MU):** Untreated or raw water that is supplied to replace losses (blowdown, leaks). It is the source of all impurities.
- **Feedwater (FW):** A chemically treated mixture of makeup water and recovered condensate. This water is conditioned to remove corrosive gases (deaeration) and preheated (economizer) before being pumped to the boiler.

- **Boiler Water (BW):** The body of water inside the boiler shell or steam drum, where the majority of heat transfer and evaporation occurs. This water is highly concentrated with non-volatile impurities.

Steam Generation and Delivery (The Transformation Phase):

- **Separation:** In recirculation boilers, a mixture of steam and water is produced in the generating tubes and flows to the Steam Drum where mechanical internal separators ensure high-quality saturated steam is collected at the top.
- **Superheating:** The saturated steam often passes through a Superheater (located in a high-temperature gas zone) to raise its temperature above the saturation point, yielding Superheated Steam. This increases the energy content and prevents premature condensation in turbines.

Condensate Return (The Recovery Phase):

- **Condensation:** After the steam expands across a turbine or delivers its latent heat to a process (e.g., a reboiler or heat exchanger), it cools and collapses back into a liquid state known as Condensate.
- **Recovery:** This pure condensate is collected via Steam Traps and returned to the feedwater section. Its high purity and temperature make its recovery essential for energy efficiency and minimizing fresh water treatment costs.

2.2 Heat Transfer Basics: Radiation, Convection, and Conduction Explained Simply

For efficient steam generation, heat must move quickly and effectively from the fuel source to the water. All three fundamental modes of heat transfer are employed within the boiler structure:

Radiation (The Direct Blaze): This is the transfer of heat via high-energy electromagnetic waves. It does not require a medium and is the primary mechanism within the furnace's immediate vicinity. The Waterwalls—the tubes lining the furnace—are designed to capture this radiant energy. Because radiant energy transfer is proportional to the fourth power of the flame's absolute temperature, it is the most intense form of heat uptake.

Conduction (The Solid Bridge): This is the molecular transfer of heat through a solid material. In a boiler tube, heat must be conducted from the outer fireside surface, through the steel of the tube wall, and across the thin layer of boiler water film to the waterside fluid. Any insulating deposit (scale or sludge) on the waterside is a severe block to conduction and can cause tube failure.

Convection (The Flowing Heat): This involves the transfer of heat due to the mass movement of a heated fluid or gas. Once the hot flue gases leave the radiant furnace, they move across the Convection Section of the boiler. The remaining thermal energy is "scrubbed" from the flue gases by forced contact with the tubes, feedwater economizers, and air heaters.

Analogy: Understanding the various heat transfer modes is like understanding how a space heater works. Heat radiates directly off the hot coils (radiation), warms the surrounding air molecules (conduction), and then the lighter warm air rises and circulates through the room (convection).

2.3 Circulation: Natural vs. Forced Circulation

Circulation is the engineered movement of water, steam, or both through the boiler's internal circuits. Its role is twofold: **to prevent metal overheating by continuously sweeping** the heat away from the tube walls, and **to distribute chemical treatment** and remove solids. Inadequate circulation is one of the quickest paths to tube failure.

2.3.1 Natural Circulation: The "Coffee Percolator" Effect

Natural Circulation (or Thermo-Siphon) relies solely on the fundamental principle that water density decreases as it is heated.

- **Downcomers:** These pipes are positioned in the cooler areas of the boiler, carrying denser, non-boiling water from the drum to the bottom headers.
- **Risers (Evaporators):** These tubes are exposed to the heat source. As water inside the risers is heated, steam bubbles form, creating a lighter, less dense two-phase (steam-water) mixture.

The column of heavy water in the downcomers creates a pressure head that continuously pushes the lighter steam-water mixture up and out of the risers and back into the steam drum. This process generates a powerful, self-sustaining flow required to cool the evaporator tubes. Natural circulation is highly reliable and is the chosen method for boilers operating up to approximately **175 kg/cm²** pressure.

2.3.2 Forced Circulation

Forced Circulation is achieved mechanically using pumps to drive the flow through the boiler circuits. It is mandatory for two main reasons:

- **High-Pressure Operation:** As boiler pressure increases toward the critical point (3,208 psia), the density difference between water and steam diminishes to zero, making natural circulation impossible. Forced circulation pumps must overcome the frictional resistance of the system to maintain flow.
- **Once-Through Designs:** Systems that lack a steam drum (like many Supercritical boilers) are inherently forced-circulation systems. All the feedwater entering the boiler leaves as steam, requiring a dedicated pump to manage the continuous flow.

2.4 Key Parameters: Pressure, Temperature, and Steaming Rate

Boiler operation is a precise exercise in managing phase change, which is quantified by three interlinked parameters.

Pressure and Temperature (The Saturation State):

- **Saturation Condition:** This is the precise pressure and temperature pairing at which liquid water and steam can coexist.
- **Sensible Heat:** The heat required to raise the water temperature from its starting point up to the saturation temperature.
- **Latent Heat:** The heat required to change saturated water into saturated steam without changing the temperature. This is the energy that does the work.
- **Superheated Steam:** Steam that is heated beyond its saturation temperature, making it "dry" and highly energetic. As boiler pressure increases, the percentage of total heat required as sensible heat increases, while the latent heat required decreases, leading to lower energy transfer capacity per pound of steam at extreme pressures.

Steaming Rate (Capacity):

- This is the output of the boiler, often expressed as the **Maximum Continuous Rating (MCR)** in pounds or kilograms of steam per hour (lb/hr or kg/hr).
- **Boiler Horsepower (BHP):** An older unit, still used, defined as the energy required to evaporate 34.5 lb/hr of water from and at 212°F. Steaming rate is the critical input for feedwater control systems, ensuring the mass flow in equals the mass flow out.

Circulation Ratio:

- This is a critical design parameter: the ratio of the total mass of water-steam mixture flowing through the generating tubes to the mass of steam actually produced.
- Example: A Circulation Ratio of 10:1 means that ten pounds of water circulate through the riser for every one pound of steam generated. The nine pounds of excess water provide the cooling necessary to protect the tube from overheating. A high circulation ratio is essential for high heat-flux areas.

2.5 Steam Quality and Purity: The Ultimate Goal

The entire purpose of the boiler is not just to produce steam, but to produce **high-quality, high-purity steam**. Steam quality and purity are two distinct, critical metrics that dictate the safety and efficiency of the plant downstream.

2.5.1 Steam Quality (The Physical State)

Steam quality is a physical measurement of the wetness or dryness of the steam.

- **Saturated Steam**: Steam that is in equilibrium with its liquid phase (water) at the boiling temperature. It is ready to condense immediately upon losing any heat. If this steam leaves the drum with water droplets entrained in it, it is known as Wet Steam.
- **Dryness Fraction (X):** This is the mass fraction of vapor in a wet steam mixture. For example, a dryness fraction of

0.98 means the steam is 98% vapor and 2% entrained water droplets. A dryness fraction of 1.0 is 100% dry, saturated vapor.

- **Superheated Steam:** Steam heated above its saturation temperature. This steam is inherently dry (Dryness Fraction = 1.0) and is preferred for power turbines and long-distance transport because it can lose a significant amount of heat before condensation (wetness) begins.

2.5.2 Steam Purity (The Chemical State)

Steam purity is a chemical measurement of the non-volatile impurities (salts, silica, sodium) carried over with the steam from the boiler water. These impurities are dangerous:

- **Turbine Deposits:** The biggest risk. Impurities, especially Silica and Sodium, vaporize and travel with the steam. When the steam expands through the turbine blades, these impurities condense and solidify onto the blades, causing imbalance, efficiency loss, and ultimately, catastrophic mechanical failure.
- **Conductivity:** The primary field measurement of steam purity is its Cation Conductivity, which is essentially a measurement of dissolved minerals and salts. Maintaining extremely low conductivity (often below $0.1 \mu\text{S}/\text{cm}$) is mandatory for high-pressure boilers.
- **Carryover:** The term for the condition where excessive boiler water (BW) and its impurities are physically swept out of the steam drum with the steam. This is usually caused by excessive water levels, high steaming rates, or chemical fouling (foaming).

Analogy: Steam Quality is like the physical dryness of a road surface (is it wet, damp, or perfectly dry?). Steam Purity is like the chemical cleanliness of the surface (is it covered in dirt, gravel, or perfectly swept?). You need both 100% dry steam and 100% clean steam for safe turbine operation.

2.6 Critical Components of the Steam Loop

While the boiler itself is the heart, the surrounding auxiliary equipment is the circulatory system—and they are essential components of the vocabulary.

2.6.1 The Deaerator (DA) – The Boiler's Lung

Corrosion is the single greatest threat to a steam system. The primary corrosive agents are dissolved oxygen (O_2) and carbon dioxide (CO_2). The **Deaerator (DA)** is the first and most vital piece of equipment dedicated to their removal.

- **Mechanism**: The DA operates on the principle that the solubility of gases in water decreases as temperature increases. Feedwater is sprayed into the DA, meeting an atmosphere of low-pressure steam. This steam heats the water to its saturation temperature, forcing the dissolved O_2 and CO_2 out of solution.
- **Function**: It is a mechanical process that reduces dissolved oxygen content from several parts per million (ppm) down to the chemical target range (often less than 7 parts per billion (ppb)), making the job of the chemical oxygen scavengers much easier.

2.6.2 The Economizer (Econ)

Located in the flue gas exit duct, the Economizer is purely an efficiency component. It is a coil of tubes that preheats the returning **Feedwater (FW)** using the waste heat from the hot flue gases before they exit the stack. By raising the feedwater temperature, less fuel is required by the main burner to convert the water to steam. Every 10°F rise in feedwater temperature via the economizer typically translates to a 1% saving in fuel consumption.

2.6.3 The Steam Trap: The Automated Gatekeeper

Steam traps are arguably the most overlooked component in the system, yet their failure can be enormously costly. A steam trap is an automatic valve designed to perform three critical functions:

1. **Discharge Condensate:** It must remove the condensate (liquid water) from the steam line as soon as it forms.
2. **Pass Steam:** It must be able to hold back steam, ensuring that only liquid is discharged.
3. **Discharge Air/Non-Condensable Gases:** It must vent non-condensable gases (like air or CO_2 which would otherwise foul the heat transfer surfaces.

If a trap fails "open," it wastes valuable steam (a Blow-Through failure). If it fails "closed," it causes the steam line to back up with corrosive condensate (a Blockage failure).

2.7 Boiler Efficiency Metrics

Measuring boiler efficiency is the process of quantifying how well the chemical energy in the fuel is converted into usable heat energy in the steam.

2.7.1 Combustion Efficiency vs. Boiler Efficiency

1. **Combustion Efficiency:** Measures only how completely the fuel is burned. This is primarily determined by the percentage of O_2 (or CO_2) in the flue gas and the stack temperature.
2. **Boiler Efficiency (Thermal Efficiency):** This is the true measure of performance. It is the ratio of the energy output (steam) to the energy input (fuel). The ASME Power Test Codes define the precise methods for calculating this efficiency, with two common methods:

- The Input-Output Method: The most precise method, relying on measuring the mass flow and enthalpy of steam and the fuel flow.
- The Heat Loss Method: The most common field method, which calculates efficiency by subtracting all quantifiable heat losses (stack gas, radiation, moisture in fuel) from 100%.

2.7.2 Efficiency Loss: The Cost of Blowdown

One unavoidable loss is the **Blowdown (BD)**. Blowdown is the controlled removal of a portion of the concentrated boiler water (BW) to limit the buildup of solids and prevent scaling and carryover.

- The Loss: The water removed is at the high temperature of the boiler, meaning its sensible heat is wasted. This heat loss is significant and proportional to the amount of blowdown required.
- Heat Recovery: To mitigate this loss, systems utilize Blowdown Heat Recovery flash tanks and heat exchangers to preheat the feedwater using the thermal energy in the discharged boiler water. This is a critical function of efficient boiler management.

2.8 Chapter Summary and Key Takeaways

- **Steam Quality:** A physical measure of wetness. For high-energy use, steam must be **Superheated** to ensure a 100 dry state (Dryness Fraction} = 1.0) upon arrival.
- **Steam Purity:** A chemical measure. Low **Cation Conductivity** is mandatory to prevent **Silica** and **Sodium** deposits on turbine blades, which cause catastrophic failure.
- **Deaerator (DA):** The key mechanical system for feedwater treatment. It heats water to saturation temperature to expel corrosive gases, primarily **Oxygen** (O_2) and **Carbon Dioxide** (CO_2).

  Economizer: An efficiency device that uses waste heat from flue gas to preheat feedwater. Every 10 °F gain saves 1% in fuel.
- **Steam Traps:** Automated valves that are critical for removing high-energy condensate from steam lines without wasting steam (preventing **Blow-Through**).
- **Boiler Efficiency:** : Measured by the ratio of energy

output (steam) to energy input (fuel). Blowdown is a necessary but costly heat loss that must be managed by **Heat Recovery** systems.

- **The Steam Loop: :** A continuous cycle of transforming liquid FW into **Steam**, recovering the heat energy, and collecting high-purity **Condensate** for reuse.
- **Heat Transfer: :** Dominated by **Radiation** in the waterwalls, supplemented by **Convection** in downstream sections. **Conduction** through the tube wall is compromised by scale.
- **Circulation: Natural Circulation** relies on the density difference (**Thermo-Siphon**). **Forced Circulation** is mandatory for supercritical or very high-pressure units.
- **Circulation Ratio:** The ratio of mass flow to steam generated; it is the most critical factor for ensuring evaporator tubes are properly cooled and preventing tube failures.
- **Saturation State:** For every boiling pressure, there is only one corresponding **Saturation Temperature**; this relationship defines the boiler's operational curve and energy potential.

3. THE HARDWARE

3.1 Firetube Boilers: The "Tank" Design

The **Firetube (FT)** boiler is the venerable workhorse of small-to-medium industry. Its design is intuitive: it is essentially a large, enclosed steel shell (the tank) filled with water. A series of tubes runs through this water, and the "fire" (hot combustion gases) travels inside those tubes.

Because the water surrounds the tubes, the outer shell must contain the full pressure of the steam. This creates a mechanical limit: to hold high pressure in a large-diameter shell, the steel would have to be impossibly thick and heavy. Consequently, FT boilers are generally limited to pressures **below 250 psi** and diameters under **10 feet.**

The Scotch Marine Style: Most modern FT boilers are "packaged" units based on the Scotch Marine design. They feature an internal furnace tube (the first pass) where the most intense radiant heat transfer occurs, followed by several banks of smaller "smoke tubes."

3.1.1 Two-Pass, Three-Pass, and Four-Pass Configurations

The efficiency of a firetube boiler is determined by how much "contact time" the hot gas has with the water. This is measured in **Passes**.

- **Two-Pass:** The gases go down the furnace and back through one set of tubes. Simple, but less efficient.
- **Three-Pass:** The most common industrial standard. The gases travel the length of the boiler three times, extracting more heat with each turn.
- **Four-Pass:** The practical limit. By the fourth pass, the gases have cooled significantly; while highly efficient, these units require more maintenance to prevent "cold-end corrosion" in the final tubes.

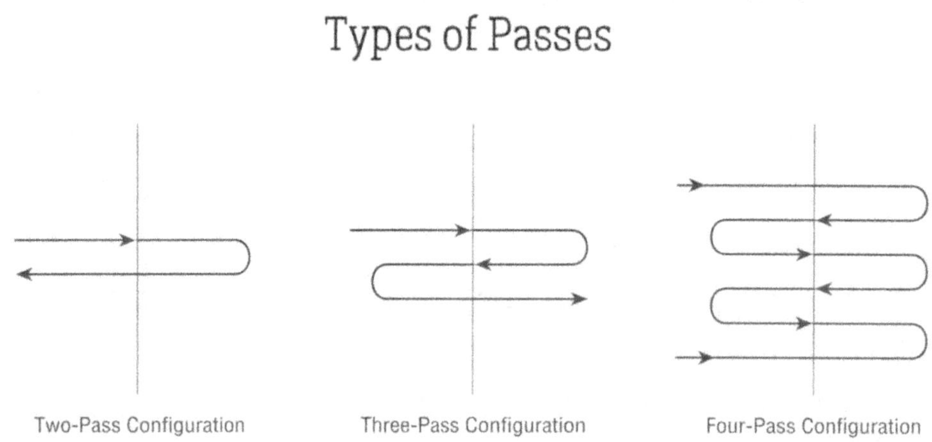

Image of Two-Pass, Three-Pass, and Four-Pass Configurations

The "Winding Road" Analogy: Imagine driving a car to cool the engine. A two-pass boiler is like a short straightaway; a four-pass boiler is like a winding mountain road. The longer the gas stays on the "road" inside the water, the more heat it gives up.

3.2 Watertube Boilers: The "Spiderweb" Design

When a plant needs massive amounts of steam or pressures exceeding 300 psi, the Firetube design reaches its limit. We then turn to the Watertube (WT) boiler. In this "Spiderweb" design, the roles are reversed: the water stays inside the small-diameter tubes, and the fire roars around the outside.

The Safety of Small Diameters: Physics dictates that it is much easier to contain high pressure in a 2-inch tube than in an 8-foot shell. Therefore, WT boilers can operate at extreme pressures (over 3.0 psi). Furthermore, because they hold much less water than a firetube "tank," they can respond almost instantly to changes in steam demand.

D-Type, O-Type, and A-Type Configurations

Packaged watertube boilers are often named after the shape the tubes form between the top **Steam Drum** and the bottom **Mud Drum**.

- **D-Type:** The most popular packaged design. The drums are offset, creating a large, flexible furnace area on one side (the "D" shape).
- **O-Type:** The drums are stacked vertically. This design is compact and symmetrical, often used where floor space is

limited, though it is harder to clean.

- **A-Type:** Features one steam drum and two smaller mud drums at the bottom. This offers a very stable, balanced weight distribution, ideal for large-scale industrial use.

3.3 The Steam Drum: The Critical Separation Zone

In a watertube boiler, the Steam Drum is the "Control Tower." It is a heavy pressure vessel at the very top of the system. It performs three vital functions:

- **Storage:** It holds a reserve of water to prevent the tubes from running dry.
- **Chemistry:** It is the point where we inject treatment chemicals and perform **Blowdown** to remove impurities.
- **Separation:** Its most critical job is to ensure that the steam leaving the boiler is "Dry."

3.3.1 Cyclones and Scrubbers: The Battle Against Carryover

As the steam-water mixture rushes into the drum from the riser tubes, it is a chaotic, turbulent mess. If we simply let the steam exit, it would carry large "slugs" of water with it—a disaster called **Carryover**.

- **Centrifugal Separators (Cyclones):** These use the "spin cycle" principle. The mixture enters tangentially, spinning the heavy water to the walls while the light steam rises through the center.
- **Scrubbers and Chevron Mist Eliminators:** These are the final "checkpoints." They consist of zig-zagging plates

or mesh pads. Steam must weave through them; any remaining microscopic water droplets hit the plates and fall back into the drum.

3.4 Burners and Furnaces: The Energy Source

The **Furnace** is the heat-release zone. In modern boilers, this isn't just a brick box; it is lined with **Membrane Waterwalls**. These are tubes welded together to form a solid wall of water. This design protects the boiler's outer casing and ensures that almost 100% of the radiant heat is absorbed by the water immediately.

The **Burner** is the "Carburetor" of the system. It must precisely mix fuel (Gas, Oil, or Coal) with Air.

- **Atomization**: For liquid fuels, the burner must "mist" the oil into tiny droplets (Atomization) using high pressure or steam.
- **BMS (Burner Management System):** This is the electronic brain that ensures the fire is safe. It manages the "Purge" (clearing out unburned gas) and monitors the "Flame Scanner" to ensure the fire never goes out unexpectedly.

3.5 Superheaters and Reheaters: The High-Energy "Add-Ons"

If the steam drum is the delivery room, the Superheater is the "finishing school." As we discussed in Chapter 2, saturated steam carries a risk: as soon as it loses even a fraction of heat, it turns back into water droplets. In a high-speed turbine, these droplets act like tiny bullets, eroding the blades.

The Hardware Solution: A Superheater is a bank of tubes located directly in the path of the hottest flue gases. Saturated steam from the drum is routed back through these tubes. Because the steam is no longer in contact with liquid water, its temperature rises far above the saturation point.

- **Radiant Superheaters:** Located in the furnace, seeing the "glow" of the fire. They absorb heat quickly but are at high risk if steam flow drops.
- **Convective Superheaters:** Located further back in the gas path, absorbing heat from the moving hot air.
- **Reheaters:** Used in large power plants, these take steam that has already passed through one stage of a turbine, send it back to the boiler to be "re-heated," and send it back to the next turbine stage for maximum efficiency.

The "Hair Dryer" Analogy: Saturated steam is like air coming off a damp towel. Superheated steam is like the air coming out of a high-performance hair dryer. It's dry, it's hot, and it carries much more energy to get the job done.

3.6 Attemperators and Desuperheaters: The Steam Thermostat

While we want hot steam, we don't want it too hot. If steam temperature exceeds the design limits of the turbine or the piping metallurgy, the metal will "creep" (permanently stretch) and eventually fail.

The Hardware: An **Attemperator (or Desuperheater)** is a device used to regulate steam temperature.

- **Direct Contact (Spray) Type:** High-purity water is sprayed directly into the superheated steam flow. The water evaporates, absorbing heat and lowering the steam temperature.
- **Surface Type:** Steam passes through a heat exchanger submerged in the boiler water.

The "Chemical Expert" Note: For spray-type attemperators, the spray water must be ultra-pure (typically polished condensate). Since this water never sees the steam drum's separation equipment, any minerals in the spray water will go directly into the turbine.

3.7 The Mud Drum: The Boiler's "Kidney"

In a Watertube boiler, the bottom drum is called the **Mud Drum**. Unlike the steam drum, this drum is completely filled with water.

The Function: Its primary purpose is to act as a settling chamber. Because the water velocity is lower here, heavy sludge and precipitated minerals (the "mud") settle to the bottom.

Bottom Blowdown: This is where the heavy-duty blowdown valves are located. By opening these valves intermittently, the operator "purges" the accumulated sludge from the system.

3.8 Essential Boiler Mountings (Safety Hardware)

No boiler is complete without the hardware mandated by the ASME Code to ensure safe operation. These are known as **Mountings**.

1. **Safety Relief Valves:** The most critical hardware. They are designed to "pop" open at a set pressure to prevent the boiler from over-pressurizing. They must be capable of discharging all the steam the boiler can produce.
2. **Water Column and Gauge Glass:** The operator's "eyes." This allows for a physical, visual verification of the water level. In the age of digital sensors, the physical gauge glass remains the final authority on water level.
3. **Pressure Gauge:** Must be visible to the operator and connected to the highest part of the steam space via a "siphon" (a pigtail pipe) to protect the gauge from direct steam contact.

3.9 Soot Blowers: Keeping the Fireside Clean

Just as scale fouls the inside of the tubes, **Soot** (ash and unburned carbon) fouls the outside. Soot acts as a powerful insulator, blocking heat transfer and driving up fuel costs.

The Hardware: Soot Blowers are long, retractable lances that use high-pressure steam or air to "blast" the soot off the tubes while the boiler is in operation. Without them, a coal or oil-fired boiler would lose efficiency within hours.

3.10 Chapter Summary and Key Takeaways

- **Superheaters** are essential for protecting turbines and increasing energy density by making steam "dry" and extra-hot.
- **Attemperators** act as the system's thermostat. Direct spray attemperators require the highest purity water in the entire plant to protect downstream equipment.
- **The Mud Drum** is the collection point for solids; it requires regular "Bottom Blowdown" to keep the boiler's "blood" clean.
- **Safety Mountings** (Safety valves and gauge glasses) are non-negotiable legal requirements under the ASME and IBR codes. They are the final line of defense against catastrophe.
- **Fireside Maintenance:** Hardware like Soot Blowers ensures that the external "skin" of the tubes remains clean, allowing for maximum heat absorption.
- **The Pass System:** In Firetube boilers, more passes (up to 4) generally mean higher efficiency but increased mechanical complexity.
- **Shape Matters:** Packaged Watertube boilers (A, D, O-types) are designed to fit specific footprints while maximizing radiant heat surface area.
- **Dry Steam is Mandatory:** Mechanical internals like Cyclones and Scrubbers in the steam drum are the only things preventing water from "carrying over" and destroying downstream turbines.
- **Membrane Walls:** Modern furnaces use water-cooled

membrane walls to increase heat absorption and reduce the need for heavy, expensive refractory brick.

Analogy for the Road: If a boiler is a car, the Burner is the engine, the Steam Drum is the transmission, and the Safety Valve is the airbag. You might drive for years without needing the airbag, but you wouldn't dream of taking the car on the highway without it. In the next chapter, we move from the hardware to the "Software"—the Water Pretreatment required to keep this iron and steel alive.

4. PRETREATMENT – THE FIRST LINE OF OPERATION

4.1 Impurities in Water: What are we fighting?

Water is often called the "**universal solvent**." While this is great for washing dishes, it is a nightmare for boilers. Raw Water (RW)—whether from a river, well, or city municipality—is never pure. It is a chemical soup of dissolved minerals, gases, and suspended solids.

If these impurities enter the boiler, they do not just "pass through." As the water boils off into pure steam, these impurities are left behind, concentrating rapidly in the drum. We fight three primary enemies:

- **Hardness (The Insulator)**: Hardness is defined by the presence of Calcium (Ca^{2+}) and Magnesium (Mg^{2+}) salts. In the high heat of a boiler, these minerals come out of solution and bake onto the hot metal tubes, forming Scale. **The Danger:** Scale acts as a perfect insulator. A layer of scale as thin as an eggshell can reduce heat transfer efficiency by over 10% and cause the tube metal to overheat, blister, and rupture.

- **Iron (The Fouler):** Iron (Fe) typically enters the boiler not from the raw water, but from the corrosion of your own condensate lines. This "transported iron" returns to the boiler and settles as a heavy sludge in low-flow areas (like the mud drum) or deposits on top of scale, creating porous layers that trap corrosive chemicals against the tube wall.
- **Silica (The Glass Maker):** Silica (SiO_2) is dissolved sand. It creates two distinct problems:

1. Hard Scale: In the boiler, silica forms a glass-hard scale that is incredibly difficult to remove (often requiring hydrofluoric acid cleaning).
2. Volatilization: At pressures above 600 psi, silica effectively vaporizes. It travels with the steam and deposits on turbine blades, destroying their aerodynamic profile and causing vibration.

4.2 Softening: The "Ion Swap"

For the vast majority of low-to-medium pressure industrial boilers, Sodium Zeolite Softening is the standard for removing hardness.

The Mechanism:

1.Softening is a chemical trade. The **softener vessel** is filled with millions of tiny plastic beads (Resin). These beads are chemically charged and coated with Sodium (Na^+) ions.

- Service Cycle: As hard water flows past the beads, the resin has a stronger magnetic attraction to Calcium and

Magnesium than it does to Sodium. It grabs the Ca^{2+} and Mg^{2+} and releases the Na^+ into the water.

Result: The dangerous calcium is trapped on the bead, and the harmless sodium goes to the boiler. Sodium is highly soluble and does not form scale.

2.**Regeneration Cycle**: Eventually, the beads are full of calcium. We must "recharge" them. We soak the beads in a strong Brine (Salt) solution. The sheer concentration of salt forces the calcium off the beads and replaces it with fresh sodium. The calcium-rich brine is then flushed down the drain.

4.3 Demineralization and RO: Removing the Salts Completely

As boiler pressures rise (typically above 600 psi), even Sodium is no longer acceptable. High-pressure boilers require water that is essentially chemically void.

Demineralization (DI): This is "Softening on Steroids." Instead of just swapping hardness for sodium, we swap everything.

- Cation Units: Swap positive ions (Ca, Mg, Na) for Hydrogen (H^+).
- Anion Units: Swap negative ions (Cl, SO_4, SiO_2) for Hydroxyl (OH^-).
- The Magic: When the H^+ meets the OH-, they combine to form H_2O. The result is ultra-pure water.

Reverse Osmosis (RO): This is a physical filtration process at the molecular level. High-pressure pumps force water against a semi-permeable membrane. The water molecules are small enough to squeeze through, but larger salt ions, organics, and silica are blocked and swept away in a "reject" stream.

- **Role:** RO is essentially the "workhorse" that removes 98% of the TDS (Total Dissolved Solids) before the water hits the polishing demineralizers.

Table 4.1 Pretreatment Parameter

Pretreatment Step	Target Impurity	Mechanism	Why is it Critical?
Softening	Hardness (Ca, Mg)	Ion Exchange (Na+ swap)	Prevents Scale, which insulates tubes and causes overheating.
Demineralization / RO	Total Salts (TDS) & Silica	Membrane / Ion Exchange (H+/OH-)	Required for high-pressure boilers to prevent carryover and turbine blade deposits.
Deaeration	Oxygen (O2) & CO2	Thermal Stripping	Prevents Pitting Corrosion (Red Death) in the boiler and piping.

4.4 Deaeration: The Mechanical Lung

Once the minerals are gone, we must deal with the gases. **Dissolved Oxygen (O_2)** is the primary cause of corrosion in the feedwater piping and the boiler itself.

The Principle: Gases dissolve easily in cold water but cannot stay dissolved in boiling water. (Think of how a soda goes flat when it gets warm).

The Deaerator (DA): The DA is a tank mounted high in the plant. It uses steam to heat the incoming feedwater to within a few degrees of saturation (212°F to 220°F).

1. **Mechanical Stripping:** By spraying the water into a steam atmosphere, the water is heated and agitated. The oxygen is physically "scrubbed" out and vented to the atmosphere.
2. **Thermal Shock Protection:** By preheating the water, the DA also prevents cold water from shocking the hot boiler metal.

4.4.1 The "Red Death": Oxygen Pitting

If the DA fails, oxygen enters the boiler. In the high-heat environment, oxygen attacks the steel tubes, drilling deep, pinhole pits. This is often called "Red Death" because the pits are capped with red iron oxide tubercules.

The Chemical Polish: A DA can remove oxygen down to 7 ppb (parts per billion). To get to zero, we add a chemical **Oxygen Scavenger** (like Sodium Sulfite or Hydrazine) to react with the final trace amounts.

4.5 Chemical Oxygen Scavenging: The "Final Polish"

While the mechanical Deaerator (DA) does the heavy lifting (removing roughly 99% of dissolved oxygen), it cannot reach zero. It typically leaves behind 7 to 10 ppb (parts per billion) of oxygen. In the high-temperature environment of a boiler, even 10 ppb is enough to cause rapid localized pitting. Therefore, we must use a chemical "mop" to scavenge the remaining oxygen.

4.5.1 Sodium Sulfite (Na2SO3): The Standard for Low Pressure

For the vast majority of industrial boilers (below 900 psi), **Sodium Sulfite** is the scavenger of choice.

- **The Reaction:** Sulfite (Na_2SO_3) reacts with Oxygen (O_2) to form Sodium Sulfate (Na_2SO_4).

$$2Na_2SO_3 + O_2 \rightarrow 2Na_2SO_4$$

- **The Pro:** It is cheap, fast-acting (especially with a cobalt catalyst), and easy to test for.
- **The Con:** It adds solids (Conductivity) to the boiler water. More critically, at pressures above 900 psi, sulfite breaks down into corrosive sulfur dioxide gas (SO_2) and hydrogen sulfide (H_2S), which can destroy the turbine. Therefore, **Sulfite is banned in high-pressure steam generation**.

4.5.2 Hydrazine (N2H4) and Organic Scavengers: The High-Pressure Solution

For high-pressure utility boilers, we use volatile scavengers that do not add solids. Hydrazine was the industry standard for decades.

- **The Reaction:** Hydrazine + Oxygen → Nitrogen + Water.

$$N_2H_4 + O_2 \rightarrow N_2 + 2H_2O$$

- **The Benefit:** The byproducts are harmless gas and water.

Zero solids. It also helps "passivate" the metal surfaces, creating a protective magnetite layer (Fe_3O_4).

- **The Risk:** Hydrazine is a known carcinogen. Because of this safety risk, many plants have switched to "Safe Substitutes" like C**arbohydrazide or DEHA (Diethylhyd roxylamine)**, which do the same job without the cancer risk.

Analogy: If the Deaerator is a vacuum cleaner sucking up the dirt (oxygen), the Chemical Scavenger is the "Sanitizing Wipe" used afterwards. The vacuum gets the big stuff, but the wipe ensures the surface is sterile.

4.6 Feedwater pH Control: Preventing "Flow-Accelerated Corrosion"

Pure water is actually aggressive to steel. If we feed ultra-pure RO or Demineralized water into carbon steel piping, it will strip the protective oxide layer off the metal, leading to a dangerous thinning of the pipes known as **Flow-Accelerated Corrosion (FAC).**

To prevent this, we must artificially elevate the pH of the feedwater to a safe range (typically **8.8 – 9.2** for mixed metallurgy systems, or **9.2 – 9.6** for all-ferrous systems).

- **Ammonia/Amines:** We inject volatile alkalizing agents (like Morpholine, Cyclohexylamine, or Ammonia) into the feedwater. These chemicals travel with the water and steam, keeping the pH high and the protective oxide layer intact throughout the cycle.

Table 4.2: Water Treatment Method Comparison

Parameter	Low Pressure (<10 bar)	Medium (10-40 bar)	High (>40 bar)
FEEDWATER			
pH	8.8-9.2	9.0-9.5	9.2-9.6
Hardness (ppm $CaCO_3$)	<2	<0.3	<0.05
Iron (ppb)	<100	<50	<10
Oxygen (ppb)	<7	<7	<3
Conductivity (µS/cm)	<10	<5	<2
BOILER WATER			
pH	10.5-11.5	10.5-11.0	9.8-10.2
Alkalinity (ppm)	300-500	200-400	150-300
Phosphate (ppm PO_4)	20-40	30-60	5-15
Sulfite (ppm SO_3)	30-50	40-60	0
Conductivity (µS/cm)	<3500	<3000	<1000
Silica (ppm SiO_2)	<150	<100	<8

4.7 Pretreatment Monitoring: The Operator's Watch

The most common cause of boiler scaling is not the lack of a softener, but a **Softener Overrun.** This happens when the resin bed is exhausted, but the operator keeps running water through it.

4.7.1 Hardness Leakage

- **The Rule:** Boiler feedwater hardness should be Non-Detectable.
- **The Test:** Operators must perform a hardness test (colorimetric titration) on the softener effluent at least once per shift.
- **The Danger Zone:** If the test turns from blue (soft) to purple/red (hard), you are feeding "liquid stone" into your boiler. Immediate Action: Switch to the standby softener and regenerate the exhausted unit.

4.7.2 Chlorine Attack on Resin

Softener resin is made of plastic (polystyrene). It is sensitive to oxidizers like Chlorine, which is found in almost all city water. Over time, chlorine turns the resin into "mush," causing pressure drops and loss of capacity.

- **The Fix**: If your raw water has high chlorine ($>0.5 \text{ ppm}$), you should install an **Activated Carbon Filter** or inject **Sodium Bisulfite** before the softener to neutralize the chlorine and protect the expensive resin.

4.8 Chapter Summary and Key Takeaways

- **Chemical Scavenging is Mandatory:** Mechanical deaeration is not enough. You must use a chemical scavenger (Sulfite for low pressure, Hydrazine/Substitutes for high pressure) to remove the final trace oxygen.
- **Feedwater needs Buffering:** Pure water eats steel. We must inject amines or ammonia to keep the Feedwater pH between 8.8 and 9.6 to prevent Flow-Accelerated Corrosion (FAC).
- **The "Softener Overrun" is the #1 Failure:** A softener only works until its resin is full of calcium. Running it past this point injects massive amounts of scale into the boiler.
- **Testing Parameter and Target:**
- **Hardness**: Test every shift. Target: Zero.
- **Sulfite Residual:** Test every shift. Target: 30–50 ppm (typical for low pressure).
- **Deaerator Temperature:** Monitor continuously. A drop in DA temperature means a spike in oxygen.

Note: A wise operator knows that **scale is permanent, but chemicals are temporary.** If you let hard water into a boiler for just 24 hours, you have deposited scale that will stay there for years, reducing efficiency every single day until you chemically clean it. The battle is won or lost in the Pretreatment Room, not the Boiler Room.

5. CHEMISTRY – THE BALANCING ACT

5.1 The Objectives: Why Do We Treat the Water?

Once the water is inside the boiler, the rules change. We are no longer just filtering or softening; we are managing a high-pressure chemical reactor. Internal treatment is the "Final Line of Defense." Even with the best pretreatment, some impurities will enter the boiler. The internal chemistry program exists to handle these breakthroughs without destroying the boiler.

The three practical objectives are:

1. **Keep Surfaces Clean (Scale Control):** Heat transfer surfaces must remain metal-clean. Any deposit acts as an insulator, leading to higher fuel bills and tube ruptures.
2. **Keep Metal Intact (Corrosion Control):** We must maintain a specific chemical environment (pH and Oxygen-free) that allows the steel to form its own protective shield—a black magnetic oxide called **Magnetite** (Fe_3O_4).
3. **Keep Steam Pure (Carryover Control):** We must prevent the "dirty" boiler water from foaming up and carrying salts into the clean steam system.

5.2 Oxygen Scavenging: The Chemical Mop

Mechanical deaerators are great, but they aren't perfect. A boiler operating at 600 psi will suffer pitting damage if oxygen levels exceed even 7 ppb. Since a good DA only gets down to 7–10 ppb, we need a chemical "polisher."

Table 5.1: Type of Oxygen Scavenging

Scavenger	Type	Best Application	Operator Notes
Sodium Sulfite (Na_2SO_3)	Inorganic Salt	Low to Medium Pressure (< **900 psig)**	**Cheap & Effective.** Easy to test. ***Downside:*** Adds solids (TDS) to the boiler, increasing blowdown requirements. Breaks down into corrosive H_2S at high pressures.
Hydrazine (N_2H_4)	Volatile	High Pressure Utility (> **1000 psig)**	**Zero Solids.** Reacts to form N_2 and water. Passivates metal well. ***Downside:*** **Carcinogenic.** Requires strict safety handling.
Carbohydrazide	Volatile	High Pressure / Safety Conscious	**Safe Hydrazine Substitute.** Does not add solids. Breaks down slowly into hydrazine inside the boiler to do the work without the handling risk.

Operator Tip: Monitor your scavenger residual daily. If your residual suddenly drops to zero, **check your Deaerator.** A mechanical failure in the DA (like a broken tray or vent) puts a massive load on the chemical scavenger that it cannot handle.

5.3 Scale Control: Two Philosophies

How do we handle Calcium that slips past the softener? We have two choices: Turn it into soft mud (Precipitation) or keep it invisible (Solubilization).

5.3.1 Phosphate Programs: The "Precipitation" Approach

This is the traditional method. We intentionally inject Phosphate (PO_4) into the boiler.

- **The Mechanism:** Calcium + Phosphate = Calcium Phosphate (Hydroxyapatite).

$$10Ca^{2+} + 6PO_4^{3-} + 2OH^{-} \rightarrow Ca_{10}(OH)_2(PO_4)_6 \text{ (Soft Sludge)}$$

- **The Goal:** We want this solid to form. But unlike sticky Calcium Carbonate scale, Calcium Phosphate is a soft, non-sticky sludge. It settles to the bottom of the Mud Drum and is easily removed via **Bottom Blowdown**.
- **Key Control:** You must maintain high Alkalinity (pH > 10.2). If pH drops, you form "Sticky" phosphates that cause scale.

5.3.2 Polymer/Chelant Programs: The "Solubilization" Approach

This is the modern method for cleaner boilers.

Chelants (e.g., EDTA): These are like chemical claws. They grab the Calcium ion and lock it into a soluble ring structure. The Calcium literally cannot react with anything else. It stays dissolved and leaves the boiler harmlessly with the Continuous Blowdown.

Risk: If you overfeed Chelants, the "claws" will start attacking the boiler metal itself (Chelant Corrosion).

Polymers (Dispersants): These are long-chain molecules that act like electrical insulators for crystals. They coat floating particles and prevent them from sticking to each other or the tube walls. They keep the dirt "dispersed" so it floats out with the blowdown.

5.4 Cycles of Concentration: The "Salty Soup" Analogy

Imagine boiling a pot of salted soup. As the water steams off, the soup gets saltier and saltier. If you don't add fresh water, eventually the salt becomes so thick it turns into a solid crust on the pot.

- The **Feedwater** is the fresh soup ingredients.
- The **Steam** is the pure water vapor leaving (taking zero salt with it).
- The **Boiler Water** is the super-salty leftover broth.

Cycles of Concentration (COC) tells us how concentrated that broth is.

COC = Boiler Water Conductivity (BW) / Feedwater Conductivity (FW)

Example: If FW is 100µS and BW is 2000 µS, you are running at 20 Cycles. This means the water in the boiler is 20 times more concentrated than the water coming in.

5.4.1 Why We Blowdown

We cannot let the Cycles go to infinity. Eventually, the dissolved solids will reach saturation and turn into scale (no matter how much chemical we add).

- **Continuous Blowdown (CBD):** A small valve near the surface of the steam drum that is always open. It skims off the concentrated water to keep the COC stable.
- **Bottom Blowdown:** A large valve at the bottom of the mud drum. Opened briefly (e.g., once per shift) to "flush the toilet"—removing the heavy sludge that has settled out of the chemistry.

5.5 Steam Purity and the Danger of Carryover

The ultimate goal of internal chemistry is not just to protect the boiler, but to ensure that the steam leaving the drum is 99.9% pure. When boiler water—along with its dissolved salts and silica—is physically carried out with the steam, we call this **Carryover**.

5.5.1 Priming vs. Foaming

Carryover is usually caused by one of two conditions:

- **Priming:** A mechanical event. This is the sudden "gulping" of water into the steam outlet. It is usually caused by operating with a water level that is too high, or a sudden, massive increase in steam demand (load swing) that causes the water to "swell" violently.
- **Foaming:** A chemical event. High concentrations of dissolved solids, organics, or high alkalinity change the surface tension of the boiler water. Instead of bubbles bursting cleanly, they build up into a thick foam that fills the steam space and is sucked into the headers.

The "Soda Bottle" Analogy: Priming is like tilting a soda bottle too far and getting a mouthful of liquid. Foaming is like shaking the bottle before opening it—the bubbles expand and overflow no matter how careful you are.

5.6 The "Big Four" Control Parameters

As an operator, you aren't just "dumping chemicals." You are managing four critical data points. If these four are in range, your boiler is likely safe.

Conductivity (TDS): Measures the "thickness" of the soup.

- Control: Adjust **Continuous Blowdown (CBD).** If conductivity is too high, open the CBD.

Alkalinity (OH or P-Alkalinity): Ensures the pH is high

enough to protect metal and facilitate phosphate reactions.

- Control: Adjust the feed rate of your **Alkalinity Builder** (typically Caustic Soda).

Phosphate / Polymer Residual: Ensures there is enough "active ingredient" to fight any hardness that enters.

- Control: Adjust the feed rate of your **Scale Inhibitor**.

Sulfite / Scavenger Residual: Ensures the water is "hungry" for oxygen.

- Control: Adjust the feed rate of your **Oxygen Scavenger**.

5.7 Chemical Feeding Strategies

Where you put the chemicals is just as important as how much you use.

- **Feedwater Line:** Oxygen scavengers and amines should be fed into the storage section of the Deaerator or the feedwater line. This protects the piping before the water reaches the boiler.
- **The Steam Drum:** Scale inhibitors (Phosphates/Chelants) are often fed directly into the steam drum via a dedicated chemical feed line. This ensures the chemical is at its highest concentration exactly where the boiling occurs.

5.8 Lay-up: Protecting the Idle Boiler

More boilers are destroyed by corrosion while they are off than while they are on. When a boiler cools down, a vacuum forms, sucking in oxygen-rich air.

- **Wet Lay-up:** For short-term standby. The boiler is filled completely to the top (including the superheater) with water treated with double the normal dose of oxygen scavenger and alkalinity.
- **Dry Lay-up:** For long-term storage. The boiler is drained and dried completely. Trays of desiccant (like silica gel) are placed inside the drums to soak up any moisture, and the boiler is sealed tight.

5.9 Chapter Summary and Key Takeaways

- **Carryover is main problem:** Whether caused by mechanical **Priming** or chemical **Foaming**, carryover introduces solids into the steam that will eventually erode valves and destroy turbines.
- **Monitoring is Mandatory:** An operator should never "fly blind." Daily testing of **Conductivity, Alkalinity, Phosphate, and Sulfite** is the minimum requirement for safe operation.
- **Continuous vs. Bottom Blowdown:** Use CBD to control dissolved solids (Conductivity). Use Bottom Blowdown to remove physical solids (Sludge/Mud).
- **Never Ignore the Idle Boiler:** Oxygen is most aggressive in a cooling, stagnant boiler. Always implement a Wet or Dry Lay-up procedure when the fire is out.

- **Scavenging: Sulfite** is for low pressure (cheap, easy, adds solids). **Hydrazine/Carbohydrazide** is for high pressure (volatile, zero solids, passivating).
- **Scale Control (Phosphate)**: turns hardness into soft sludge (requires bottom blowdown).
- **Scale Control (Chelant/Polymer):** keeps hardness dissolved or dispersed (cleaner tubes, but risk of corrosion if overfed).
- **Cycles of Concentration (COC):** The measure of how "thick" the boiler water is compared to feedwater. High cycles save water/heat but increase the risk of scale.
- **Blowdown:** The mechanical limit. We must bleed off concentrated water (CBD) and settled sludge (Bottom Blowdown) to keep the chemical balance safe.

Note: You can't manage what you don't measure. A water treatment program is only a "program" if there is a logbook showing consistent testing and adjustments. If the tests are out of range, the chemistry isn't working—it's just an expensive way to pollute your blowdown drain.

6. OPERATION AND CONTROL

6.1 Startup Curves: Why Slow and Steady Wins the Race

A cold boiler is a stressed boiler. The transition from ambient temperature to operating pressure is the most dangerous time for the mechanical integrity of the unit. We call this "**Waking the Beast.**"

The Enemy: Thermal Shock

Steel expands when heated. However, thick metal (like the Steam Drum walls, often 2–5 inches thick) heats up slower than thin metal (the tubes).

- **The Stress:** If you fire too hard, the tubes expand rapidly while the drum is still cold. This differential expansion creates massive tensile stress at the rolled joints and welds.
- **The Result:** Micro-cracking. These cracks may not fail today, but they will grow with every startup cycle until catastrophic rupture occurs.

The Golden Rule: 100°F Per Hour Standard industry practice dictates that the water temperature should not increase by more than 100 °F (55 °C) per hour.

- **Intermittent Firing:** For firetube boilers, this means firing the burner for 1 minute and leaving it off for 5 minutes. You must "soak" the heat into the metal.
- **The Refractory Factor:** Refractory brick holds moisture. If you heat faster than the water can evaporate from the brick, the trapped steam will blow the face off the refractory (Spalling).

The Critical Vent Valve: The drum vent valve must remain OPEN until the boiler pressure reaches 25 psig.

1. **Oxygen Removal:** We must drive out all the air (oxygen) to prevent pitting.
2. **Vacuum Prevention:** On shutdown, the vent must be opened *before* the pressure hits zero. If not, the condensing steam creates a vacuum that can suck gaskets in or even collapse the boiler shell.

6.2 Water Level Control: The Most Critical Safety Loop

The boiler water level is the heartbeat of the system.

- **Too Low:** You starve the tubes. Without water to cool them, the steel reaches plastic deformation temperature (>1000 °F) in seconds and ruptures. **Low water is the #1 cause of boiler explosions.**
- **Too High:** You flood the steam separators. Water slugs carry over into the steam line, hammering check valves and destroying turbine blades.

6.2.1 Shrink and Swell: Why the Water Level Lies

Water level is not static; it is dynamic. The level you see in the glass is a mixture of water and steam bubbles.

- **Swell (The Fake High):**
- *Scenario:* A sudden demand for steam (e.g., a process valve opens).
- *Physics:* The drum pressure drops. This causes the steam bubbles submerged in the water to expand rapidly.
- *Result:* The water level physically rises, even though the boiler is losing mass.
- *Danger:* A simple controller will close the feedwater valve to lower the level—exactly when the boiler needs *more* water.
- **Shrink (The Fake Low):**
- *Scenario:* Sudden load rejection (steam demand stops).
- *Physics:* Drum pressure spikes. The bubbles are compressed and collapse.
- *Result:* The level collapses (shrinks), causing the controller to flood the boiler.

6.2.2 The Solution: Multi-Element Control

To defeat Shrink and Swell, we stop looking at just the level.

- **Single-Element:** Looks only at Drum Level. (Okay for steady loads).
- **Two-Element:** Looks at Drum Level + **Steam Flow**. (If steam leaves, add water *immediately*, regardless of what the level says).

- **Three-Element:** Looks at Drum Level + Steam Flow + **Feedwater Flow**. (The industry standard for power boilers).

6.3 Blowdown Strategies: Continuous vs. Intermittent

We must remove the concentrated "soup" (Conductivity) and the settled "mud" (Sludge). These require two different valves.

6.3.1 Continuous Blowdown (Surface Blow)

- **Location:** The "Skimmer" pipe, just below the water surface in the steam drum.
- **Purpose:** To remove dissolved solids (TDS) and silica. This is where the concentration is highest.
- **Operation:** Adjusted automatically based on conductivity. It runs 24/7.
- **Efficiency:** This hot water is usually routed to a **Flash Tank** to recover low-pressure steam for the Deaerator.

6.3.2 Bottom Blowdown (Mud Blow)

- **Location:** The very bottom of the mud drum or firetube shell.
- **Purpose:** To remove heavy sludge, rust, and scale that has settled out of gravity.
- **The "Circulation Killer":** Opening this valve disrupts the natural thermo-siphon circulation. **NEVER** open the bottom blowdown while the burner is on high fire. You can starve a tube and melt it.
- **The Safety Sequence:** Bottom blowdown lines usually

have two valves: a Quick-Opening valve (Sealing) and a Slow-Opening valve (Blowing).

- **To Open:** Open Quick → Open Slow.
- **To Close:** Close Slow → Close Quick.
- *Why?* This protects the Quick valve from wire-drawing (erosion) by the high-velocity water.

6.4 Layup Procedures: Protecting the Idle Boiler

Corrosion doesn't stop when the fire goes out; it accelerates. An idle boiler sucks in oxygen as it cools. You must choose a strategy based on time.

Table 6.1: Parameter Maintenance on Idle Boiler

Method	Duration	Procedure	Chemical Targets
Wet Layup	Short Term (< 30 Days)	Fill boiler completely to the vent (eliminate air space). Keep pressurized with N_2 or a head tank.	**Sulfite:** 200 ppm **pH:** 11.0 (High alkalinity prevents pitting)
Dry Layup	Long Term (> 30 Days)	Drain while hot (>180 °F) so it flash-dries. Place desiccant trays (Silica Gel) inside drums. Seal perfectly.	**Humidity:** < 30 % inside the drum. **Visual:** Change desiccant when it turns pink.

Operator Warning: Never enter a boiler that has been in Nitrogen Blanketing layup without thoroughly venting it first.

Nitrogen is odorless and causes asphyxiation in seconds.

6.5 Combustion Efficiency: The Art of "Excess Air"

Running a boiler isn't just about safety; it's about money. The largest controllable cost in the plant is fuel. The operator controls this cost by managing the **Air-to-Fuel Ratio**.

6.5.1 The "Goldilocks" Principle

To burn fuel completely, you need oxygen.

- **Too Little Air (Fuel Rich):** You create unburned fuel (Soot) and Carbon Monoxide (CO). This is dangerous (explosion risk) and inefficient.
- **Too Much Air (Lean):** You are heating up thousands of cubic feet of extra nitrogen and blowing it out the stack. This is literally "burning money."

The Target: Excess Air

We cannot mix air and fuel perfectly, so we always feed a little extra air to be safe. This is called Excess Air.

- **Natural Gas Targets:** Typically **10% to 15% Excess Air** (roughly 2–3% O_2 in the stack).
- **Oil Targets:** Typically **15% to 20% Excess Air** (roughly 3–4% O_2 in the stack).

Operator Action: Watch your Stack Temperature and Flue Gas Oxygen analyzer. If your O_2 climbs from 3% to 6%, your efficiency has dropped significantly. Check your burner linkages and damper settings.

6.6 Routine Safety Testing: The Operator's Ritual

You cannot trust a safety device unless you test it. The ASME code and insurance inspectors require specific tests to be performed by the operator.

6.6.1 The Low-Water Fuel Cutoff (LWCO) Test

This is the most critical test in the boiler room. It proves that if the water drops, the fire will stop.

- **The "Slow Drain" Test:**

1. With the burner firing, slowly open the drain valve on the LWCO water column.
2. Watch the water level in the glass drop.
3. **Result:** The burner should shut down *before* the water leaves the visible part of the glass.

- **The "Evaporation" Test (The True Test):**

1. Manually shut off the feedwater pump.
2. Allow the boiler to steam off the water naturally.
3. **Result:** The burner must trip automatically as the level hits the setpoint.
4. *Note:* This is a high-stress test and should only be done under strict supervision, but it verifies the *entire* loop (float chamber is not clogged).

6.6.2 The Flame Scanner Test

How do you know the scanner actually sees the fire?

- **Procedure:** With the boiler running, pull the UV scanner out of its mount (or cover the eye).
- **Result:** The Burner Management System (BMS) should instantly trip the fuel valves (Main Flame Failure). If it takes more than 2–4 seconds, the system is faulty.

6.7 Emergency Procedures: Responding to a "Low Water" Alarms

One day, the alarm **will** go off. How the operator reacts in the next 60 seconds determines if the plant shuts down safely or if the boiler explodes.

The Golden Rule: NEVER add water to a hot, dry boiler.

Scenario: The Low Water alarm sounds, and you look at the gauge glass.

1. **If you can SEE water in the glass:**

- It is a "controlled" low event.
- **Action:** Manually start the feedwater pump or bypass the control valve to restore level. Troubleshoot why the automatic fill failed.

1. **If you CANNOT see water (Glass is Empty):**

- **Action: TRIP THE BURNER IMMEDIATELY.** (If it hasn't already).

- **CRITICAL STEP: DO NOT ADD WATER.**
- *Why?* You do not know how low the water is. The tubes may be red-hot (1,200 °F). If you inject 200 °F feedwater, the thermal shock will instantly flash the water to steam and rupture the boiler casing.
- **Recovery:** Isolate the boiler. Allow it to cool naturally (slowly) for several hours. Only then, open the inspection hatches to look for melted tubes or warped plates.

Table 6.2 Quick Troubleshooting Guide

Problem	Potential Cause	Corrective Action
High Stack Temp	Soot on tubes (Fireside) Scale on tubes (Waterside)	Run Soot Blowers. Inspect for Scale (Schedule cleaning).
Low Stack Temp	Low firing rate Excess air too high	Check combustion controls. Tune burner (Oxygen trim).
Water Level Bouncing	High Solids (Foaming) High Alkalinity	Open Surface Blowdown immediately. Check chemical feed.
Low Feedwater Temp	Deaerator malfunction Heat exchanger fouling	Check DA steam valve. Check Economizer bypass.
High Conductivity	Blowdown valve clogged Controller failure	Blow down manually. Calibrate probe.
Low pH (Acidic)	Chemical pump failure Condensate contamination	Check Amine/Caustic pump. Test condensate for process leaks.

6.8 Chapter Summary and Key Takeaways

- **Startup:** Limit heating to 100 °F/hr. Keep vents open until 25 psig to remove air.
- **Level Control:** Use **Three-Element Control** (Steam Flow + Feed Flow + Drum Level) to defeat Shrink/Swell logic errors.

- **Combustion:** Maintain Excess Air targets (typically 10–15%) to balance safety with fuel efficiency. High excess air wastes heat; low excess air creates dangerous CO.
- **Testing:** Perform the **LWCO Slow Drain test** daily/weekly. Perform the **Evaporation Test** monthly/quarterly to prove the float chamber is clear of mud.
- **Emergency Response:** If the gauge glass is empty, **Kill the Fire and Do Not Add Water.** Adding water to a dry boiler is the fastest way to cause an explosion.
- **Thermal Shock:** Limit heating rates to **100 °F per hour** to prevent cracking thick metal components. Keep the vent open until **25 psig**.
- **Shrink & Swell:** Rapid load changes cause false level readings. Use **Two-Element** or **Three-Element** control to feed water based on steam flow, not just drum level.
- **Blowdown Discipline:**
- **Surface (Continuous):** For Chemistry/TDS control.
- **Bottom (Intermittent):** For Sludge control. Only done at low/no fire. Use the "Quick-Slow-Slow-Quick" valve sequence.
- **Safety:** Bottom blowdown removes water so fast it can reverse circulation and overhead tubes if the boiler is steaming heavily.
- **Layup is Mandatory:** An idle boiler without chemical protection is a rusting boiler. Use **Wet Layup** for standby and **Dry Layup** for seasonal shutdowns.

Note: An operator who knows *how* to start a boiler is good. An operator who knows *when to shut it down* (and refuses to override a safety limit) is a professional. Your license is a certificate of trust—don't violate it by bypassing a safety switch "just to keep production running."

7. TROUBLESHOOTING – READING THE CLUES

7.1 Introduction: The Boiler is Talking to You

A boiler rarely fails "suddenly." In 90% of cases, a catastrophic failure is the final act of a long, slow process of chemical imbalance or thermal stress. The boiler tube tells a story. By examining the location, shape, and texture of a failure, a skilled operator can determine exactly what went wrong weeks or months ago.

Troubleshooting is not just fixing the leak; it is performing an autopsy to ensure it doesn't happen to the next tube.

7.2 Scale and Deposits: The "Insulator"

As discussed in Chapter 2, heat transfer relies on **Conduction** through the tube metal. Scale is a deposit of minerals (Calcium, Magnesium, Silica) that forms a barrier on the water side of the tube.

The Mechanism: Scale acts as an insulator. It has a thermal conductivity roughly 50 times lower than steel.

1. The fire heats the outside of the tube.
2. The scale blocks the heat from reaching the water.
3. The water cannot cool the tube metal.
4. The metal temperature skyrockets (often exceeding $1,000^\circ\text{F}$).
5. The steel loses its tensile strength, bulges out like a balloon, and eventually bursts.

Visual Clues:

- **Blisters:** Visible bumps on the fireside of the tube.
- **"Fish Mouth" Rupture:** A wide, gaping hole with **thin edges** (Thin-Lip Rupture). The metal stretched like taffy before breaking because it was soft and hot.

Analogy: Imagine trying to boil water in a paper cup. As long as the water touches the paper, the paper won't burn because the water cools it. Now, imagine putting a piece of cardboard inside the cup (Scale). The heat hits the paper, can't get to the water, and the paper burns instantly.

7.3 Corrosion Mechanisms: The Chemical Attack

While scale causes overheating, corrosion causes metal loss. The type of hole tells you the type of chemical failure.

7.3.1 Oxygen Pitting: The "Drilled Hole"

- **The Cause:** Failure of the Deaerator or Oxygen Scavenger (Chapter 4).
- **The Appearance:** Deep, round pits with steep sides. They often look like someone took a drill press to the tube.
- **The tell-tale sign:** The pits are often covered by a "cap" or "tubercle" of hard, red/black iron oxide. If you pop the cap off, you find active black liquid corrosion underneath.
- **Location:** Most common in the Feedwater Line, Economizer, and the Steam Drum water line.

7.3.2 Caustic Gouging: The "Chemical Burn"

- **The Cause:** High pH (Caustic) concentrating under a layer of porous sludge. This is a side effect of "Hideout."
- **The Mechanism:** Water boils off beneath a pile of dirt, leaving a super-concentrated caustic solution (pH 14+) against the metal. This dissolves the protective magnetite layer.
- **The Appearance:** Irregular, smooth grooves or gouges in the metal. It looks like the metal has been scooped out with a spoon.
- **Location:** High heat transfer zones (furnace floor/walls) where sludge settles.

7.3.3 Acid Attack: The "Orange Peel"

- **The Cause:** Low pH (Acidic water). This can happen if regeneration acid from a demineralizer leaks into the feedwater, or if the pH control fails.

- **The Appearance:** A general thinning of the metal with a rough, etched surface texture similar to an orange peel.

7.4 Carryover and Steam Purity Issues

When the "separation" in the steam drum fails, boiler water travels with the steam.

Troubleshooting Signs:

1. **Superheater Failure:** The superheater is designed for dry gas. If water hits it, the solids dry out inside the U-bends, blocking flow. The tube overheats and fails.
2. **Turbine Vibration:** Silica deposits on turbine blades create imbalance.
3. **Water Hammer:** Loud banging in the steam lines caused by slugs of water moving at 60 mph.

The Root Causes:

- **Mechanical:** Damaged cyclone separators in the drum.
- **Operational:** Running the water level too high.
- **Chemical:** Foaming due to high Alkalinity, high Suspended Solids, or oil contamination.

7.5 Tube Failure Identification: The Autopsy

When a tube bursts, the shape of the hole is the best evidence.

7.5.1 Thin-Lip Rupture (Short-Term Overheating)

- **Scenario:** You ran the boiler low on water (Low Water Cutoff failed). The cooling stopped immediately.
- **Look For:** The edges of the burst are **razor sharp** (knife-edged). The tube swelled significantly before popping.
- **Diagnosis:** This happened fast. The metal was extremely hot (1,400+ °F) and ductile.

7.5.2 Thick-Lip Rupture (Long-Term Overheating)

- **Scenario:** Heavy scale buildup (Concept 7.2). The tube ran slightly hotter than normal for months.
- **Look For:** The edges of the burst are **thick, rough, and blunt**. There is very little swelling.
- **Diagnosis:** This is "Creep Failure." The metal structure slowly degraded over time until it cracked.

7.5.3 Fatigue Failure (The Stress Crack)

- **Scenario:** Cycling the boiler on/off too quickly (Thermal Shock).
- **Look For:** Transverse cracks (running around the circumference of the tube). They are straight, not jagged.
- **Diagnosis:** The tube expanded and contracted too many times, like bending a paperclip back and forth until it snaps.

7.6 Operational Troubleshooting: When the Equipment Fails

Tube failures are catastrophic, but equipment failures are the daily headaches. Here are the three most common non-pressure part failures an operator will face.

7.6.1 Feedwater Pump Cavitation: The "Gravel" Sound

The Symptom: The boiler feedwater pump sounds like it is pumping gravel or marbles. Discharge pressure fluctuates violently.

The Cause: Cavitation. This occurs when the water temperature in the Deaerator is too high for the available pressure at the pump suction (insufficient Net Positive Suction Head or NPSH).

- **The Physics:** The hot water literally boils *inside* the pump impeller. The bubbles form in the low-pressure eye and violently collapse (implode) when they hit the high-pressure vanes.
- **The Fix:**

1. Check the Deaerator level (is it too low?).
2. Check the Deaerator pressure (did it drop suddenly, allowing the water to flash?).
3. Check the suction strainer (is it clogged?).

7.6.2 Flame Failure / Scanner Faults

The Symptom: The boiler trips on "Main Flame Failure," but you are sure there was a fire.

The Cause: The Burner Management System (BMS) lost sight of the flame.

- **Dirty "Eye":** The UV scanner lens is coated in soot or oil. Wipe it clean.
- **Sighting:** The scanner is looking at the black refractory instead of the root of the flame.
- **Fuel/Air Ratio:** The fire is too lean (lifting off the burner tip) or too rich (smoky and obscuring the scanner).

7.6.3 Safety Valve "Simmering"

The Symptom: The safety valve isn't popping, but it is hissing or weeping steam.

The Cause:

- **Operating too close to setpoint:** You cannot operate a 150 psig boiler at 148 psig. The valve spring needs a differential to seal. Keep operating pressure at least 10% below the safety valve setpoint.
- **Debris:** A piece of scale or wire is stuck on the valve seat.
- **WARNING:** Never try to "gag" or tighten a leaking safety valve. You must shut down the boiler and have a certified valve technician repair it.

7.6.4 Feedwater Pump Cavitation: The "Gravel" Sound

The Symptom: The boiler feedwater pump sounds like it is pumping gravel or marbles. Discharge pressure fluctuates violently. **The Cause: Cavitation.** This occurs when the water temperature in the Deaerator is too high for the available pressure at the pump suction (insufficient Net Positive Suction Head or NPSH).

- **The Physics:** The hot water literally boils *inside* the pump impeller. The bubbles form in the low-pressure eye and violently collapse (implode) when they hit the high-pressure vanes.
- **The Fix:**
- Check the Deaerator level (is it too low?).
- Check the Deaerator pressure (did it drop suddenly, allowing the water to flash?).
- Check the suction strainer (is it clogged?).

7.6.5 Flame Failure / Scanner Faults

The Symptom: The boiler trips on "Main Flame Failure," but you are sure there was a fire. **The Cause:** The Burner Management System (BMS) lost sight of the flame.

- **Dirty "Eye":** The UV scanner lens is coated in soot or oil. Wipe it clean.
- **Sighting:** The scanner is looking at the black refractory instead of the root of the flame.
- **Fuel/Air Ratio:** The fire is too lean (lifting off the burner tip) or too rich (smoky and obscuring the scanner).

7.6.6 Safety Valve "Simmering"

- **The Symptom:** The safety valve isn't popping, but it is hissing or weeping steam. **The Cause:**
- **Operating too close to setpoint:** You cannot operate a 150 psig boiler at 148 psig. The valve spring needs a differential to seal. Keep operating pressure at least 10% below the safety valve setpoint.
- **Debris:** A piece of scale or wire is stuck on the valve seat.
- **WARNING:** Never try to "gag" or tighten a leaking safety valve. You must shut down the boiler and have a certified valve technician repair it.

7.7 Chapter Summary and Key Takeaways

- **Scale = Insulation:** Scale blocks heat, causing the tube temperature to rise until it fails. Look for blisters.
- **Oxygen Pitting:** Identified by deep, round holes often covered by red oxide caps. It indicates a failure in pretreatment (DA) or scavenger feed.
- **Caustic Gouging:** Occurs *under* sludge deposits. The best prevention is keeping the boiler clean (Bottom Blowdown) to prevent sludge accumulation.
- **Lip Thickness Matters:**
- **Thin Lip (Sharp Edge):** Sudden loss of water (Acute Overheating).
- **Thick Lip (Blunt Edge):** Long-term scale buildup (Chronic Overheating/Creep).
- **Fatigue:** Cracks caused by rapid startup/shutdown cycles (Thermal Shock).

8. EFFICIENCY AND OPTIMIZATION

8.1 The Real Cost of Steam: The "Money Burning Machine"

A boiler is likely the most expensive piece of equipment to operate in your entire facility. The annual fuel cost for a standard industrial boiler can easily exceed **five times its initial purchase price**.

The Efficiency Equation:

Boiler efficiency is simply the ratio of useful energy output to energy input.

$$\text{Efficiency}(\%) = (\text{Steam Energy Output}) * 100 / (\text{Fuel Energy Input})$$

- **Combustion Efficiency:** How well the fuel burns (typically > 80%).
- **Thermal Efficiency:** How much of that heat actually gets into the water (typically 75%–85% for older units, > 95% for condensing units).

The "leaky bucket" Analogy: Imagine carrying water in a

bucket with holes. The water at the destination is your steam. The water leaking out the holes is your heat loss. You can pour faster (burn more fuel), or you can plug the holes (optimize efficiency).

8.2 Stack Loss: The #1 Enemy

The largest "hole in the bucket" is the chimney. Every degree of temperature leaving the stack is energy that *didn't* go into the water.

- **The Rule of Thumb:** For every **40 °F(22 °C)** you reduce the stack temperature, you gain **1% efficiency**.
- **The Dew Point Floor:** You cannot cool the gas too much. If flue gas drops below its dew point (typically 250°F–300°F for sulfur-bearing fuels), corrosive acid condenses and eats the stack.

8.3 Heat Recovery Hardware: Plugging the Holes

To capture that escaping heat, we use specific hardware "add-ons."

8.3.1 The Economizer

As mentioned in the Pretreatment chapter, the Economizer is a heat exchanger placed in the exhaust stack.

- **Function:** It uses hot exhaust gas to preheat the boiler **Feedwater**.
- **The Benefit:** Since the water enters the boiler already hot,

the burner doesn't have to work as hard to boil it.

- **Limit:** You must ensure the feedwater entering the economizer is not *too* cold (usually kept $>220^\circ\text{F}$ via the deaerator) to prevent "cold-end corrosion" from condensing acid on the tubes.

8.3.2 The Air Preheater

Used primarily in large utility boilers.

- **Function:** Uses exhaust gas to preheat the **Combustion Air** going into the burner.
- **The Benefit:** Hot air helps fuel ignite faster and burn more completely.

8.4 Blowdown Heat Recovery: Don't Drain the Cash

We established in Chapter 5 that Blowdown is mandatory to control chemistry. But blowdown water is **350°F+ treated water**. Dumping it directly to the drain is a double loss: you lose the heat energy *and* the water/chemical cost.

The Solution: A Two-Stage Recovery System

1. **Flash Tank:** The high-pressure blowdown enters a low-pressure tank. About 10–15% of the water instantly flashes into low-pressure steam. This clean steam is sent to the **Deaerator** to strip oxygen.
2. **Heat Exchanger:** The remaining hot liquid passes through a plate-and-frame heat exchanger to warm up the cold **Makeup Water**.

- **Result:** You recover up to 90% of the energy from the blowdown stream.

8.5 Variable Frequency Drives (VFDs): The "Gas Pedal"

Old boilers controlled air flow using **Dampers** (louvers). The fan ran at 100% speed (full RPM) all the time, and the damper choked off the air like a hand over a vacuum cleaner. This wastes massive amounts of electricity.

The VFD Solution:

A VFD acts like a dimmer switch for the fan motor. If the boiler is at 50% load, the VFD slows the fan down to 50% speed.

- **The Physics:** The "Fan Laws" state that power consumption drops by the **cube** of the speed. Running a fan at 80% speed uses only **50% of the power**.
- **ROI:** VFD retrofits are often the quickest Return on Investment in the boiler room.

8.6 Environmental Compliance: The "Green" Constraints

Modern operators must manage not just steam, but **Emissions**. Regulatory bodies (like the EPA or local environmental agencies) strictly limit what comes out of the stack.

8.6.1 NOx (Nitrogen Oxides)

NOx causes smog and acid rain. It is created when Nitrogen in the air (or fuel) cooks at very high flame temperatures (>2,800°F).

- **Thermal NOx:** Created by the heat of the flame.
- **Fuel NOx:** Created by nitrogen chemically bound in the fuel (common in coal/oil).
- **The Fix: Low-NOx Burners (LNB).** These burners stage the fuel and air mixing to create a larger, cooler flame. By spreading the flame out, they lower the peak temperature and reduce NOx formation.
- *Trade-off:* Low-NOx burners are less stable and harder to tune. They often require tighter "Excess Air" control.

8.6.2 CO (Carbon Monoxide)

CO is the sign of **Incomplete Combustion**.

- **Cause:** Not enough air, or poor mixing of fuel and air.
- **Danger:** It is explosive and deadly.
- **Efficiency:** High CO means you are wasting unburned fuel.

8.7 Steam Trap Management: The Silent Thieves

As discussed in Chapter 2, steam traps are valves designed to remove condensate but hold back steam. However, they are mechanical devices with moving parts, and they fail.

The Failure Rate: In an unmaintained plant, **15% to 20%** of steam traps may be failed.

- **Failed Open:** The trap acts like an open pipe, blowing live steam directly into the condensate return. This is the most expensive failure mode. A single 1/8" orifice leak at 100 psig wastes roughly **$3,000 to $5,000 per year**.

- **Failed Closed:** Condensate backs up, leading to water hammer and process stalling (cold heat exchangers).

The Optimization Strategy:

You must implement a **Trap Survey** program (at least annually).

1. **Visual:** Look for "flash steam" clouds at receiver vents.
2. **Thermal:** Use an infrared gun. There should be a distinct temperature drop across the trap. If the inlet and outlet are the same temperature, it is likely blowing through.
3. **Ultrasonic:** The most reliable method. A working trap cycles (click... swoosh... silence). A failed trap hisses continuously.

> ***Analogy:*** *A failed steam trap is exactly like a toilet flapper valve stuck open. You might not see the water leaking, but you will definitely hear it, and you will definitely pay for it on the utility bill.*

8.8 Radiant Heat Loss: Insulation and Cladding

The boiler shell is massive. Even with refractory, the outer skin gets hot. Heat radiating from the boiler shell into the boiler room is energy that did not go into the steam.

The "Touch Test":

If you cannot hold your hand on the outer casing or the steam piping insulation for 5 seconds ($< 140^\circ\text{F}$), you are losing money.

- **Valves and Flanges:** These are often left uninsulated

because they are hard to wrap. Removable **Insulation Jackets** are the solution.

- **The ROI:** Insulating a single 6-inch gate valve at 150 psig saves the energy equivalent of burning 500 gallons of oil per year.

8.9 The Cost of Scale: Quantifying the Chemistry

In Chapter 4 and 5, we fought scale to prevent tube failure. In this chapter, we fight scale to save money. Scale is an insulator. When it forms on the tubes, you must burn more fuel to push the heat through the "stone" and into the water.

The Efficiency Penalty:

Even a paper-thin layer of scale has a massive impact on fuel consumption.

Table 8.1: The Efficiency Penalty

Scale Thickness	Approx. Fuel Loss
1/32 inch (0.8 mm)	7% Increase
1/16 inch (1.6 mm)	15% Increase
1/8 inch (3.2 mm)	25% Increase

Operator Insight: *If your stack temperature is slowly rising over weeks (e.g., from 350°F to 400°F) while firing at the same load, your boiler is likely scaling up. The heat is going up the chimney because it can't get into the water.*

8.10 Case Study: The "Optimized" Boiler Room

To visualize how these changes stack up, let's look at a hypothetical 500 HP firetube boiler running 24/7.

Before Optimization:

- Stack Temp: 450°F (High Excess Air, Soot on tubes).
- Blowdown: Manual valve cracked open (High heat loss).
- Feedwater: 180 °F (No Economizer).
- Steam Traps: 10% failed open.
- **Result:** Efficiency = 76%. Annual Fuel Bill = $800,000.

After Optimization:

1. **Burner Tune-up:** Reduced excess air. Stack drops to 420°F.
2. **Soot/Scale Clean:** Stack drops to 350°F. (+2.5% Eff).
3. **Economizer Install:** Feedwater rises to 230°F. (+5% Eff).
4. **Auto-Blowdown:** Reduced waste water. (+2% Eff).
5. **Trap Survey:** Fixed leaks. Saved $25,000/yr.

- **Result:** Efficiency = 85.5%. Annual Fuel Bill = $710,000.
- **Savings: $90,000 per year.** (Pays for the upgrades in < 18 months).

8.11 Chapter Summary and Key Takeaways

- **Steam Traps:** A failed trap is a steam leak. Testing via **Ultrasound** is required to identify "blow-through" failures.
- **Insulation:** Uninsulated valves and flanges act as radiators, venting heat to the room. Use removable jackets.
- **Scale Costs Money:** Just 1/16th of an inch of scale increases fuel consumption by **15%**. A rising stack temperature is the first warning sign.
- **The Holistic Approach:** Efficiency isn't one device; it is the sum of Combustion (Burner), Heat Transfer (Clean Tubes), Recovery (Economizers), and Distribution (Traps).
- **Efficiency is Economic Survival:** Fuel is the largest cost. A 1% efficiency gain can save tens of thousands of dollars a year.
- **Stack Temperature Rules:** Every 40 °F reduction in stack temp = 1% efficiency gain.
- **Recover Everything:** Use **Economizers** to catch stack heat and **Blowdown Heat Recovery** to catch water heat.
- **VFDs Save Power:** Controlling fan speed is far more efficient than choking flow with dampers.

The Emission Balance: Reducing NOx often involves lowering flame temperature, which can make the flame less stable. The modern operator must walk the tightrope between **Clean Air (Low NOx)** and **Safe Fire (Low CO)**.

9. THE OPERATOR'S EXAM – 40 CRITICAL QUESTIONS & ANSWERS

9.1 General Safety & Legal

1. What is the primary purpose of a boiler safety valve?

- To prevent the boiler pressure from exceeding its Maximum Allowable Working Pressure (MAWP). It is the last line of mechanical defense against an explosion.

2. Can you install a shut-off valve between the boiler and the safety valve?

- **NEVER.** The path between the steam drum and the safety valve must be completely unobstructed. Installing a valve here is a criminal offense in most jurisdictions.

3. What is "Purging" and why must it be done before lighting a fire?

- Purging is running the burner fan (without fuel) to clear the furnace of any combustible gases that may have accumulated. This prevents a furnace explosion upon ignition. The standard purge is 4 air changes or roughly 30–60 seconds.

4. What is the most common cause of boiler explosions today?

- **Low Water.** Firing a boiler that has no water in it leads to overheating and catastrophic metal rupture.

5. What should you do if the Low Water alarm sounds and the gauge glass is empty?

- **Shut off the fuel immediately.** Do **NOT** add water. Isolate the boiler and allow it to cool slowly. Adding water to a hot, dry boiler will cause it to flash instantly to steam, likely rupturing the shell.

6. How often should you test the Low Water Cutoff (LWCO)?

- The "Slow Drain" test should be performed daily or weekly (depending on plant policy). The "Evaporation" test (actual low water condition) should be performed monthly or quarterly.

7. What is a "Furnace Explosion" vs. a "Pressure Part Explosion"?

- **Furnace Explosion:** Ignition of accumulated fuel/gas

inside the firebox (breach of the fireside).

- **Pressure Part Explosion:** Rupture of the steel shell/tubes due to overpressure or overheating (breach of the water-side).

9.2 Operations & Controls

8. What is "Shrink and Swell"?

It is the false movement of the water level caused by pressure changes.

- **Swell:** Sudden steam demand drops pressure $\rightarrow$ bubbles expand $\rightarrow$ level rises.
- **Shrink:** Sudden load drop spikes pressure $\rightarrow$ bubbles collapse $\rightarrow$ level falls.

9. Why do we need "Continuous Blowdown"?

- To remove dissolved solids (TDS) from the steam drum. If we don't, the concentration will rise until the water foams and carries over.

10. Why do we need "Bottom Blowdown"?

- To remove heavy sludge and mud that settles in the bottom of the boiler. This must be done intermittently.

11. When is the best time to perform a Bottom Blowdown?

- When the boiler is under light load or the burner is off. This ensures the sludge has settled and minimizes the disruption to natural circulation.

12. What is the proper valve sequence for Bottom Blowdown?

"Quick-Slow-Slow-Quick."

1. Open Quick-Opening Valve (Sealing valve).
2. Open Slow-Opening Valve (Blowing valve).
3. Close Slow-Opening Valve.
4. Close Quick-Opening Valve.
5. *Reason:* This protects the expensive Quick-Opening valve from erosion (wire drawing).

13. What is "Thermal Shock"?

- Stress caused by heating or cooling the boiler metal too fast. The rule of thumb is to limit temperature changes to 100°F per hour.

14. What is the purpose of the "Vent Valve" on startup?

- To release air from the drum. It should remain open until the boiler reaches approximately 15–25 psig.

15. What is "Carryover"?

- The condition where liquid boiler water leaves the drum with the steam. It contains salts that damage valves and turbines.

16. What is the difference between "Priming" and "Foam-

ing"?

- **Priming:** Mechanical carryover caused by high water level or sudden load swing (slugs of water).
- **Foaming:** Chemical carryover caused by high TDS, oils, or high alkalinity (bubbles don't burst).

9.3 Water Treatment & Chemistry

17. Why is oxygen dangerous in a boiler?

- Dissolved oxygen causes rapid, localized pitting corrosion (drilled holes) in the tubes.

18. What device removes most of the oxygen?

- The **Deaerator (DA)**. It uses steam to heat the water to saturation (227°F at 5 psig), scrubbing out the non-condensable gases.

19. What chemical removes the *remaining* oxygen?

- An Oxygen Scavenger. **Sodium Sulfite** is used for low pressure; **Hydrazine** or Carbohydrazide is used for high pressure.

20. What is pH, and where should it be in the boiler?

- pH measures acidity/alkalinity. Boiler water should be alkaline (typically **10.5 – 11.5**) to protect the metal and allow phosphate sludge to form.

21. Why do we add Phosphate to the boiler?

- To react with Calcium hardness. It turns "sticky" scale (Calcium Carbonate) into "soft" sludge (Calcium Phosphate) that can be blown down.

22. What is "Hardness Leakage"?

- When the water softener fails and allows Calcium/Magnesium into the system. It causes rapid scaling.

23. What is "Condensate Return" and why is it valuable?

- It is the steam that has condensed back into water. It is valuable because it is already hot (180°F+) and chemically pure (distilled).

24. What causes "Carbonic Acid" corrosion in condensate lines?

- Carbon Dioxide (CO_2) in the steam dissolves into the condensate to form Carbonic Acid (H_2CO_3). This eats the bottom of condensate pipes (grooving). We use Amines to neutralize it.

25. What is the "Cycles of Concentration" (COC)?

- The ratio of solids in the boiler water to solids in the feedwater.
- COC = (Boiler Conductivity) / (Feedwater Conductivity). High cycles mean you are saving water; too high means

you risk scale.

9.4 Maintenance & Troubleshooting

26. How do you identify an "Oxygen Pitting" tube failure?

- The hole looks like a steep-walled pit or drilled hole, often covered by a red/black oxide scab.

27. How do you identify a "Short-Term Overheating" failure?

- The tube bulges and bursts with a **"Thin, Knife-Edged Lip."** This indicates the metal was extremely hot and soft when it failed (Low Water).

28. How do you identify a "Long-Term Overheating" failure?

- The tube has a thick layer of scale inside. The rupture has a **"Thick, Blunt Lip."** This is creep failure.

29. What causes a Safety Valve to "simmer" or leak?

- Operating too close to the setpoint pressure, or debris (scale/wire) trapped on the valve seat.

30. What is "Flame Impingement"?

- When the burner flame physically touches the tubes or furnace wall. This causes localized overheating and rapid failure.

31. What is "Cavitation" in a feedwater pump?

- When the water in the pump suction turns to steam (flashes) because the pressure is too low or temperature is too high. It sounds like pumping gravel and destroys the impeller.

32. How do you store a boiler for a long shutdown (Dry Layup)?

- Drain it hot, dry it out completely, place desiccant (silica gel) trays inside, and seal it tight.

9.5 Efficiency & Economics

33. What is the biggest source of energy loss in a boiler?

- **Stack Loss.** Heat escaping out the chimney.

34. What is an "Economizer"?

- A heat exchanger in the stack that uses exhaust gas to preheat the feedwater.

35. What is the rule of thumb for Stack Temperature and Efficiency?

- For every **40°F** reduction in stack temperature, you gain approx. **1% efficiency**.

36. How does a Steam Trap work?

- It distinguishes between steam and condensate (using density, temperature, or thermodynamics). It opens to let water out and closes to keep steam in.

37. How do you test a Steam Trap?

- The best way is **Ultrasonic** testing. A good trap cycles (click… flow… silence). A bad trap has a continuous rushing sound.

38. What is "Excess Air"?

- The extra air supplied to the burner (beyond what is theoretically needed) to ensure complete combustion. Too much excess air wastes heat; too little creates dangerous CO.

39. What is the visual difference between a lean fire and a rich fire?

- **Lean (High Air):** Bright white/blue, short, energetic flame.
- **Rich (Low Air):** Orange/Red, lazy, smoky flame with dark tips.

40. Why is Scale called an "Insulator"?

- It blocks heat transfer. Even 1/16th of an inch of scale can increase fuel consumption by 15% and cause tube overheating.

II

APPENDIX

APPENDIX A: STEAM TABLES

Absolute Pressure	Boiling Point	Specific Volume (steam)	Density (steam)	Specific Enthalpy of Liquid Water (sensible heat)		Specific Enthalpy of Steam (total heat)		Latent heat of Vaporization		Specific Heat
(bar)	*(°C)*	*(m³/kg)*	*(kg/m³)*	*(kJ/kg)*	*(kcal/kg)*	*(kJ/kg)*	*(kcal/kg)*	*(kJ/kg)*	*(kcal/kg)*	*(kJ/kg K)*
0.02	17.51	67.006	0.015	73.45	17.54	2533.64	605.15	2460.19	587.61	1.8644
0.03	24.10	45.667	0.022	101.00	24.12	2545.64	608.02	2444.65	583.89	1.8694
0.04	28.98	34.802	0.029	121.41	29.00	2554.51	610.13	2433.10	581.14	1.8736
0.05	32.90	28.194	0.035	137.77	32.91	2561.59	611.83	2423.82	578.92	1.8774
0.06	36.18	23.741	0.042	151.50	36.19	2567.51	613.24	2416.01	577.05	1.8808
0.07	39.02	20.531	0.049	163.38	39.02	2572.62	614.46	2409.24	575.44	1.8840
0.08	41.53	18.105	0.055	173.87	41.53	2577.11	615.53	2403.25	574.01	1.8871
0.09	43.79	16.204	0.062	183.28	43.78	2581.14	616.49	2397.85	572.72	1.8899
0.1	45.83	14.675	0.068	191.84	45.82	2584.78	617.36	2392.94	571.54	1.8927
0.2	60.09	7.650	0.131	251.46	60.06	2609.86	623.35	2358.40	563.30	1.9156
0.3	69.13	5.229	0.191	289.31	69.10	2625.43	627.07	2336.13	557.97	1.9343
0.4	75.89	3.993	0.250	317.65	75.87	2636.88	629.81	2319.23	553.94	1.9506
0.5	81.35	3.240	0.309	340.57	81.34	2645.99	631.98	2305.42	550.64	1.9654
0.6	85.95	2.732	0.366	359.93	85.97	2653.57	633.79	2293.64	547.83	1.9790
0.7	89.96	2.365	0.423	376.77	89.99	2660.07	635.35	2283.30	545.36	1.9919
0.8	93.51	2.087	0.479	391.73	93.56	2665.77	636.71	2274.05	543.15	2.0040
0.9	96.71	1.869	0.535	405.21	96.78	2670.85	637.92	2265.65	541.14	2.0156
1 [1)]	99.63	1.694	0.590	417.51	99.72	2675.43	639.02	2257.92	539.30	2.0267
1.1	102.32	1.549	0.645	428.84	102.43	2679.61	640.01	2250.76	537.59	2.0373
1.2	104.81	1.428	0.700	439.36	104.94	2683.44	640.93	2244.08	535.99	2.0476
1.3	107.13	1.325	0.755	449.19	107.29	2686.98	641.77	2237.79	534.49	2.0576
1.4	109.32	1.236	0.809	458.42	109.49	2690.28	642.56	2231.86	533.07	2.0673
1.5	111.37	1.159	0.863	467.13	111.57	2693.36	643.30	2226.23	531.73	2.0768
1.6	113.32	1.091	0.916	475.38	113.54	2696.25	643.99	2220.87	530.45	2.0860
1.7	115.17	1.031	0.970	483.22	115.42	2698.97	644.64	2215.75	529.22	2.0950
1.8	116.93	0.977	1.023	490.70	117.20	2701.54	645.25	2210.84	528.05	2.1037
1.9	118.62	0.929	1.076	497.85	118.91	2703.98	645.83	2206.13	526.92	2.1124

APPENDIX A: STEAM TABLES

Absolute Pressure	Boiling Point	Specific Volume (steam)	Density (steam)	Specific Enthalpy of Liquid Water (sensible heat)		Specific Enthalpy of Steam (total heat)		Latent heat of Vaporization		Specific Heat
(bar)	*(°C)*	*(m^3/kg)*	*(kg/m^3)*	*(kJ/kg)*	*(kcal/kg)*	*(kJ/kg)*	*(kcal/kg)*	*(kJ/kg)*	*(kcal/kg)*	*(kJ/kg K)*
2	120.23	0.885	1.129	504.71	120.55	2706.29	646.39	2201.59	525.84	2.1208
2.2	123.27	0.810	1.235	517.63	123.63	2710.60	647.42	2192.98	523.78	2.1372
2.4	126.09	0.746	1.340	529.64	126.50	2714.55	648.36	2184.91	521.86	2.1531
2.6	128.73	0.693	1.444	540.88	129.19	2718.17	649.22	2177.30	520.04	2.1685
2.8	131.20	0.646	1.548	551.45	131.71	2721.54	650.03	2170.08	518.32	2.1835
3	133.54	0.606	1.651	561.44	134.10	2724.66	650.77	2163.22	516.68	2.1981
3.5	138.87	0.524	1.908	584.28	139.55	2731.63	652.44	2147.35	512.89	2.2331
4	143.63	0.462	2.163	604.68	144.43	2737.63	653.87	2132.95	509.45	2.2664
4.5	147.92	0.414	2.417	623.17	148.84	2742.88	655.13	2119.71	506.29	2.2983
5	151.85	0.375	2.669	640.12	152.89	2747.54	656.24	2107.42	503.35	2.3289
5.5	155.47	0.342	2.920	655.81	156.64	2751.70	657.23	2095.90	500.60	2.3585
6	158.84	0.315	3.170	670.43	160.13	2755.46	658.13	2085.03	498.00	2.3873
6.5	161.99	0.292	3.419	684.14	163.40	2758.87	658.94	2074.73	495.54	2.4152
7	164.96	0.273	3.667	697.07	166.49	2761.98	659.69	2064.92	493.20	2.4424
7.5	167.76	0.255	3.915	709.30	169.41	2764.84	660.37	2055.53	490.96	2.4690
8	170.42	0.240	4.162	720.94	172.19	2767.46	661.00	2046.53	488.80	2.4951
8.5	172.94	0.227	4.409	732.03	174.84	2769.89	661.58	2037.86	486.73	2.5206
9	175.36	0.215	4.655	742.64	177.38	2772.13	662.11	2029.49	484.74	2.5456
9.5	177.67	0.204	4.901	752.82	179.81	2774.22	662.61	2021.40	482.80	2.5702
10	179.88	0.194	5.147	762.60	182.14	2776.16	663.07	2013.56	480.93	2.5944
11	184.06	0.177	5.638	781.11	186.57	2779.66	663.91	1998.55	477.35	2.6418
12	187.96	0.163	6.127	798.42	190.70	2782.73	664.64	1984.31	473.94	2.6878
13	191.60	0.151	6.617	814.68	194.58	2785.42	665.29	1970.73	470.70	2.7327
14	195.04	0.141	7.106	830.05	198.26	2787.79	665.85	1957.73	467.60	2.7767
15	198.28	0.132	7.596	844.64	201.74	2789.88	666.35	1945.24	464.61	2.8197
16	201.37	0.124	8.085	858.54	205.06	2791.73	666.79	1933.19	461.74	2.8620
17	204.30	0.117	8.575	871.82	208.23	2793.37	667.18	1921.55	458.95	2.9036

Absolute Pressure	Boiling Point	Specific Volume (steam)	Density (steam)	Specific Enthalpy of Liquid Water (sensible heat)		Specific Enthalpy of Steam (total heat)		Latent heat of Vaporization		Specific Heat
(bar)	*(^{o}C)*	*(m^3/kg)*	*(kg/m^3)*	*(kJ/kg)*	*(kcal/kg)*	*(kJ/kg)*	*(kcal/kg)*	*(kJ/kg)*	*(kcal/kg)*	*(kJ/kg K)*
18	207.11	0.110	9.065	884.55	211.27	2794.81	667.53	1910.27	456.26	2.9445
19	209.79	0.105	9.556	896.78	214.19	2796.09	667.83	1899.31	453.64	2.9849
20	212.37	0.100	10.047	908.56	217.01	2797.21	668.10	1888.65	451.10	3.0248
21	214.85	0.095	10.539	919.93	219.72	2798.18	668.33	1878.25	448.61	3.0643
22	217.24	0.091	11.032	930.92	222.35	2799.03	668.54	1868.11	446.19	3.1034
23	219.55	0.087	11.525	941.57	224.89	2799.77	668.71	1858.20	443.82	3.1421
24	221.78	0.083	12.020	951.90	227.36	2800.39	668.86	1848.49	441.50	3.1805
25	223.94	0.080	12.515	961.93	229.75	2800.91	668.99	1838.98	439.23	3.2187
26	226.03	0.077	13.012	971.69	232.08	2801.35	669.09	1829.66	437.01	3.2567
27	228.06	0.074	13.509	981.19	234.35	2801.69	669.17	1820.50	434.82	3.2944
28	230.04	0.071	14.008	990.46	236.57	2801.96	669.24	1811.50	432.67	3.3320
29	231.96	0.069	14.508	999.50	238.73	2802.15	669.28	1802.65	430.56	3.3695
30	233.84	0.067	15.009	1008.33	240.84	2802.27	669.31	1793.94	428.48	3.4069

Saturated Steam - Properties vs. Pressure - Bar

[1)] *1 bar abs = 0 bar gauge = 0.98692 atm = 100 kPa abs = Standard Atmospheric Pressure (IUPAC system).*

APPENDIX B: CONVERSION FACTORS

Look, I know unit conversions can be confusing—especially when switching between metric, imperial, and the various systems we use in India. Here's a quick reference guide I keep handy.

PRESSURE (The Most Confusing Category!)

You'll see pressure measured five different ways:

- 1 bar = 100 kPa (kilopascals)
- 1 bar = 14.5 psi (pounds per square inch)
- 1 bar $\approx$ 1 kg/cm^2 (close enough, actually 0.981 bar)
- 1 atm = 1.013 bar (atmospheric pressure)

Pro tip: When someone says "10 kg pressure," they mean 10 kg/cm^2, which is roughly 10 bar or 145 psi.

TEMPERATURE

To convert Celsius to Fahrenheit: (°C × 9/5) + 32

To convert Fahrenheit to Celsius: (°F - 32) × 5/9

Quick reference:

- Water boils: 100°C = 212°F (at 1 bar)
- Steam at 10 bar: ~180°C = 356°F

- Superheat: 250-300°C = 482-572°F

ENERGY

- 1 kcal = 4.187 kJ = 3.97 BTU
- 1 kWh = 3,600 kJ = 860 kcal

Why this matters: When fuel suppliers quote prices in kcal/kg and your efficiency is in %, you need these conversions to calculate actual costs in rupees per hour.

FLOW RATE

- 1 TPH (ton per hour) = 1,000 kg/h = 2,200 lb/h
- 1 m^3/h water = 1,000 liters/h ≈ 1 ton/h

POWER

- 1 HP (horsepower) = 0.746 kW
- 1 kW = 1.34 HP

So a 50 HP motor draws about 37 kW.

APPENDIX C: WATER CHEMISTRY TARGET TABLES

Parameter	Low Pressure (<10 bar)	Medium (10-40 bar)	High (>40 bar)
FEEDWATER			
pH	8.8-9.2	9.0-9.5	9.2-9.6
Hardness (ppm CaCO3)	<2	<0.3	<0.05
Iron (ppb)	<100	<50	<10
Oxygen (ppb)	<7	<7	<3
Conductivity (µS/cm)	<10	<5	<2
BOILER WATER			
pH	10.5-11.5	10.5-11.0	9.8-10.2
Alkalinity (ppm)	300-500	200-400	150-300
Phosphate (ppm PO4)	20-40	30-60	5-15
Sulfite (ppm SO3)	30-50	40-60	0
Conductivity (µS/cm)	<3500	<3000	<1000
Silica (ppm SiO2)	<150	<100	<8

APPENDIX D: GLOSSARY OF TERMS

- **ASME:** American Society of Mechanical Engineers - develops boiler codes
- **Attemperator**: Device to reduce superheat by injecting feedwater spray
- **Baghouse**: Fabric filter for ash/particulate removal from flue gas
- **BFW**: Boiler Feed Water - treated water supplied to the boiler
- **Blowdown**: Controlled discharge of concentrated boiler water to remove dissolved solids
- **BMS**: Burner Management System - safety control system for fuel and combustion
- **Carryover**: Water droplets carried with steam from the boiler
- **Caustic**: Alkaline chemical, typically sodium hydroxide (NaOH)
- **COC**: Cycles of Concentration - ratio of TDS in boiler water to feedwater
- Condensate: Water formed when steam condenses after giving up its heat

- **DA:** Deaerator - equipment that removes dissolved oxygen from feedwater
- **Downcomer**: Tubes carrying cooler water down in natural circulation boilers
- **Economizer**: Heat exchanger that preheats feedwater using waste heat from flue gas
- **ESP**: Electrostatic Precipitator - device for removing fly ash from flue gas
- **Feedwater**: Treated water being supplied to the boiler
- **Foaming**: Unstable boiling with excessive bubble formation at water surface
- **FT Boiler:** Fire-Tube boiler - hot gases inside tubes, water outside
- **IBR**: Indian Boiler Regulations - statutory rules for boilers in India
- **LWCO**: Low Water Cutoff - safety device that shuts burner if water level drops
- **MAWP**: Maximum Allowable Working Pressure - design pressure limit
- **Priming**: Violent boiling causing water to surge and carry over with steam
- **Riser**: Tubes carrying hot water/steam mixture up in natural circulation
- **Scale**: Hard mineral deposits on boiler tubes that reduce heat transfer
- **TDS**: Total Dissolved Solids - measure of dissolved minerals in water
- **TPH**: Tons Per Hour - common unit for steam generation rate (1 TPH = 1,000 kg/h)
- **WT Boiler**: Water-Tube boiler - water inside tubes, hot gases outside

APPENDIX E: BOILER OPERATOR CHECKLIST

BEFORE STARTING (Pre-Startup Checks):

Visual Inspection:

- ☐ Walk around boiler - look for leaks, damage, unusual conditions
- ☐ Check furnace interior - no debris, burner nozzle clean
- ☐ Inspect refractory - no cracks or damage
- ☐ Check all gauges - undamaged, readable
- ☐ Verify water level sight glass - clean, visible

Safety Devices:

- ☐ Test low water cutoff (LWCO) - drain down method or electronic test
- ☐ Check safety valve - clear, no leaks, tag current
- ☐ Verify all pressure switches - functioning
- ☐ Test flame detector - functional
- ☐ Check emergency stop button - accessible

Water Side:

- ☐ Check feedwater tank level - adequate for startup
- ☐ Verify chemical dosing pumps - operating
- ☐ Test water level controls - functional
- ☐ Check blowdown valves - closed before startup

Fuel System:

- ☐ Check fuel supply - adequate pressure/level
- ☐ Inspect fuel lines - no leaks
- ☐ For oil: Verify heating system working, temperature correct
- ☐ For gas: Check gas pressure, regulators set correctly

Air/Draft System:

- ☐ FD fan functional - no unusual sounds
- ☐ ID fan functional (if applicable)
- ☐ Dampers move freely
- ☐ Air filter clean (if applicable)

DURING OPERATION (Hourly Checks):

Operating Parameters:

- ☐ Steam pressure - record (should be stable at setpoint)
- ☐ Water level - record (middle of sight glass)
- ☐ Feedwater temperature - record (80-105°C typical)
- ☐ Stack temperature - record (<250°C typical)
- ☐ Flame appearance - bright, stable, centered

Water Chemistry (every 2-4 hours):

- ☐ pH - record (target 10.5-11.5)
- ☐ Phosphate - record (if applicable)
- ☐ Conductivity/TDS - record
- ☐ Chloride test (if checking for condenser leaks)

Blowdown:

- ☐ Bottom blowdown - perform per schedule (typically 2-4 times/shift)
- ☐ Surface blowdown - continuous or per schedule
- ☐ Check blowdown discharge - observe color, foaming

Safety:

- ☐ Safety valve area clear - no obstructions
- ☐ Emergency access clear
- ☐ Fire extinguishers accessible
- ☐ Eye wash/safety shower functional (weekly)

END OF SHIFT:

Documentation:

- ☐ Complete logbook entries
- ☐ Record all test results
- ☐ Note any abnormal observations
- ☐ Report any issues to next shift/supervisor

Hand-off:

- ☐ Brief incoming operator on current status
- ☐ Communicate any ongoing issues
- ☐ Confirm any special instructions

SHUTDOWN CHECKLIST (When Required):

Controlled Shutdown:

- ☐ Reduce firing rate gradually (don't shock the boiler)
- ☐ Continue feedwater and blowdown until pressure drops to 2-3 bar
- ☐ Shut off fuel supply
- ☐ Allow natural cooling - don't add cold water
- ☐ Maintain slight steam pressure during cooling (prevents vacuum)
- ☐ Close all valves when cool

Documentation:

- ☐ Record reason for shutdown
- ☐ Note final parameters
- ☐ Log time of shutdown

WEEKLY CHECKS (In Addition to Daily):

- ☐ Test all alarms - low water, high pressure, flame failure
- ☐ Inspect burner assembly - clean if needed
- ☐ Check all gaskets and packing - replace if leaking
- ☐ Lubricate motors and bearings
- ☐ Check flame scanner - clean lens if needed
- ☐ Test emergency shutdown system

- ☐ Review and calibrate instruments if needed
- ☐ Clean strainers and filters

MONTHLY CHECKS:

- ☐ Clean soot from tubes (if coal/oil fired)
- ☐ Inspect refractory thoroughly
- ☐ Test safety valve (typically required by IBR)
- ☐ Check expansion joints
- ☐ Inspect steam traps - replace faulty ones
- ☐ Review water treatment program effectiveness
- ☐ Calibrate instruments
- ☐ Test emergency lighting and communications

CRITICAL REMINDERS:

- ⚠ NEVER bypass the low water cutoff
- ⚠ NEVER add cold water to a hot, low-water boiler
- ⚠ NEVER enter the boiler without proper lockout/tagout
- ⚠ NEVER ignore unusual sounds, smells, or observations
- ⚠ ALWAYS maintain the logbook accurately
- ⚠ ALWAYS test safety devices regularly
- ⚠ ALWAYS follow proper startup and shutdown procedures

Remember: A safe boiler room is a quiet boiler room. If something seems wrong, investigate immediately. When in doubt, shut down safely.

About the Author

Venkatesha Perumal Ramasamy is a Chemical Engineer and Senior Sales Engineer with over 10 years of hands-on experience in industrial boilers, power plants, water treatment, and wastewater treatment systems.

He has successfully commissioned more than 5 power plants from initial installation to full operation, and worked in over 7 operational facilities across power generation, chemical, pharmaceutical, and sugar industries throughout India.

His unique combination of engineering background, commissioning experience, and operational troubleshooting gives him a 360-degree perspective on what makes boiler systems succeed or fail. He's been the engineer called at 2 AM when things go wrong—and he's learned from every midnight emergency call.

Based in Chennai, Tamil Nadu, Venkatesha wrote this book to share the practical knowledge he gained from years of real-world experience—the kind of insights textbooks often miss but operators desperately need.

His mission: Help the next generation of boiler professionals operate safely, efficiently, and with confidence.

www.ingramcontent.com/pod-product-compliance
Lightning Source LLC
LaVergne TN
LVHW082345200926
848481LV00012B/3461